Food Manipulation Technology

Wang and Hirai draw on their extensive experience to introduce recent technological advances in soft robotics for food manipulation and discuss their practical applications in the food preparation industry. They also describe food modeling and categorization for the purpose of robotic handling.

Food manipulation is common in restaurants, kitchens, and food factories, but it currently largely relies on humans because of the lack of effective robotic systems. This book therefore explains recent manipulation technology for food handling and restaurant/kitchen automation, which is essential for food manipulation. The book covers the theoretical background of soft robotic hands; their application to different handling purposes in the food industry; their use in kitchen and restaurant automation; and their use for food recognition, property measuring, and modeling. The authors also present case studies of the development of soft robotic hands to illustrate the information and provide practical information that will be instructional for readers. Readers will gain an in-depth appreciation of the technology and its practical application to industry, which will enable them to implement this in their own work.

This book is an essential read for both food technology practitioners and graduate students who are interested in food manipulation and its related technologies. It gives readers the theoretical background knowledge and practical understanding to apply this cutting-edge technology in real-world settings.

Zhongkui Wang is an Associate Professor in the Department of Robotics at Ritsumeikan University. He obtained his Bachelor's and Master's from Northeast Forestry University, Harbin, China, and his Doctorate from Ritsumeikan University in 2000, 2003, and 2011, respectively. He is a member of the Japan Society of Mechanical Engineers, Robotics Society of Japan, Japanese Society for Medical and Biological Engineering, and IEEE.

Shinichi Hirai is a Professor in the Department of Robotics and Head of the Soft Robotics Lab at Ritsumeikan University. He obtained his Bachelor's, Master's, and Doctorate Degrees from Kyoto University in 1985, 1987, and 1990, respectively. Professor Hirai is a member of the Robotics Society of Japan, Japan Society of Mechanical Engineers, Society of Instrument and Control Engineers, and IEEE Robotics and Automation Society.

Food Manipulation Technology

A Soft Robotics Approach

Zhongkui Wang and Shinichi Hirai

CRC Press
Taylor & Francis Group
Boca Raton London New York

CRC Press is an imprint of the
Taylor & Francis Group, an **informa** business

Designed cover image: @Shutterstock

First edition published 2025
by CRC Press
2385 NW Executive Center Drive, Suite 320, Boca Raton FL 33431

and by CRC Press
4 Park Square, Milton Park, Abingdon, Oxon, OX14 4RN

CRC Press is an imprint of Taylor & Francis Group, LLC

ISBN: 978-1-032-44493-2 (hbk)
ISBN: 978-1-032-44492-5 (pbk)
ISBN: 978-1-003-37243-1 (ebk)

DOI: 10.1201/9781003372431

Typeset in Latin Modern Roman font
by KnowledgeWorks Global Ltd.

Contents

List of Figures

List of Tables

About the Authors

Zhongkui Wang is an Associate Professor in the Department of Robotics at Ritsumeikan University. He obtained his Bachelor's and Master's from Northeast Forestry University, Harbin, China, and his Doctorate from Ritsumeikan University in 2000, 2003, and 2011, respectively. He is a member of the Japan Society of Mechanical Engineers, Robotics Society of Japan, Japanese Society for Medical and Biological Engineering, and IEEE.

Shinichi Hirai is a Professor in the Department of Robotics and Head of the Soft Robotics Lab at Ritsumeikan University. He obtained his Bachelor's, Master's, and Doctorate Degrees from Kyoto University in 1985, 1987, and 1990, respectively. Professor Hirai is a member of the Robotics Society of Japan, Japan Society of Mechanical Engineers, Society of Instrument and Control Engineers, and IEEE Robotics and Automation Society.

1

Introduction

1.1 Overview of Food Industry Automation

The food industry and agriculture encompass a broad range of industrial activities, including harvesting, production, processing, packaging, distribution, preparation, preservation, and service. Traditionally regarded as low-tech sectors, both the food industry and agriculture have become increasingly technology-intensive in recent years, as evidenced by their rising research and development (R&D) to sales ratio [1]. In response to challenges such as an aging population and labor shortages in countries like Japan, there has been a growing demand for automation within the food industry to maintain profit margins. Notably, the impact of the COVID-19 pandemic has further accelerated the push for automation across all stages of food production, with a particular emphasis on enhancing food safety and ensuring the stability of food supply chains [2].

In the food industry and agriculture, many operations – such as harvesting, food production, processing, and packaging – require direct contact with food products. For production and processing tasks, specialized machines – such as rice-making machines, dumpling makers, and automatic chocolate molding machines – are commonly used, as these operations are often continuous and do not require frequent adjustments. In such dedicated machines, the handling of food materials or products is typically defined by specific tasks, and pick-and-place operations are usually unnecessary.

In contrast, harvesting and packaging typically rely on pick-and-place operations. These operations are also essential for transferring food products between dedicated machines, facilitating the connection of different stages of processing. To implement pick-and-place tasks, robotic systems – including robotic manipulators, end-effectors, and sensors (e.g., cameras) – are frequently employed due to their efficiency and adaptability to a variety of food products and processing environments.

Currently, industrial robotic arms are employed as manipulators in the food industry and agriculture to perform pick-and-place operations, generating desired motions and handling payloads [3]. While industrial robotic arms offer high positional accuracy and rapid motion speeds, they are often associated with high costs. Regarding robotic end-effectors, the most commonly used type in food processing is the suction cup. Suction cups are inexpensive,

DOI: 10.1201/9781003372431-1

FIGURE 1.1
Factory scenarios of food handling operations in the food and agricultural industries: (a) bento packaging, (b) spaghetti packaging, (c) shrimp packaging, and (d) radish alignment.

simple to operate, and well-suited for many food handling tasks. Simplicity of operation is a critical requirement for food handling systems, as most tasks must be completed within a short time frame to maintain production efficiency.

However, certain food products or handling tasks – such as grasping food materials with moisture or porous surfaces – cannot be effectively managed with suction cups. Consequently, these tasks often remain dependent on human labor, as illustrated in Figure 1.1. The automation of these tasks has become a key focus for both industry professionals and researchers in the food sector.

To automate the pick-and-place operations illustrated in Figure 1.1, gripping-type robotic hands are required, and these systems must be adaptable to the variability of food products, given the frequent changes and updates in the types of food being handled. Furthermore, in many scenarios, cooked and prepared food products are randomly distributed or stored in containers rather than being neatly aligned on a conveyor belt. This variability introduces significant challenges in grasping, recognition, and sensing.

In response to these challenges, numerous studies have been conducted, and a variety of commercialized systems are now available. However, several key issues remain unresolved, presenting ongoing challenges that must be addressed in order to further accelerate automation in the food industry.

1.2 Challenges in Food Industry Automation

The development of automated systems for use in the food industry presents numerous challenges. This book specifically examines robotic systems designed to perform typical handling tasks, such as pick-and-place operations. While these tasks may appear straightforward for human workers, they pose significant difficulties for robotic systems in achieving efficiency comparable to that of human labor. Several technologies are required to successfully complete these operations, including systems for accurate food item recognition, effective handling of food products, and the design of robotic end-effectors capable of manipulating a diverse range of food items.

Food products are typically non-rigid, often fragile, and highly susceptible to bruising or marking when in contact with hard surfaces. Furthermore, they are prone to bacterial contamination, and their properties can be significantly influenced by environmental factors such as temperature, humidity, and pressure [4]. These inherent characteristics present considerable challenges in the development of robotic systems designed to handle food products.

1.2.1 Robotic end-effectors

To address the wide variety and variable characteristics of food products, the development of specialized robotic end-effectors is essential. The absence of effective end-effectors is often cited as a primary obstacle to the rapid adoption of robotic systems in the food industry [4]. Lien [5] and Fantoni *et al.* [6] summarized the challenges associated with developing robotic grippers for food handling tasks, emphasizing that robotic end-effectors must accommodate the softness, uneven surfaces, and non-uniform shapes of food products while also meeting stringent hygienic requirements. Additionally, these end-effectors must be capable of adapting to food surface conditions, such as wet or sticky surfaces. In scenarios where food products are tightly packed, robotic end-effectors must be designed to maneuver into narrow spaces for grasping. From a contamination control perspective, end-effectors should be designed with as few mechanical components as possible to minimize the risk of contamination. Moreover, from a system performance standpoint, robotic end-effectors must enable simple, high-speed motion to meet the required takt time. Finally, they should be cost-effective and adhere to hygienic design principles [7].

1.2.2 Food recognition

Challenges in food recognition vary considerably depending on operational contexts. In food processing facilities, these contexts can generally be categorized into two primary scenarios: (1) food items aligned or scattered on a conveyor belt with no overlap, and (2) food items randomly distributed within a container, often with overlapping and contact between items, a situation commonly referred to as the random bin-picking (RBP) scenario.

In the first scenario, food items or materials are positioned separately on a flat surface, making it feasible to identify them using conventional 2D image processing methods, such as color-based detection or pattern-matching techniques. In these methods, only the 2D coordinates and orientation of the food items need to be determined. Automated robotic systems are frequently employed in food factories under such conditions; for instance, robotic systems for pizza handling or powder packaging often utilize pattern-matching techniques [8].

In contrast, food recognition within the RBP scenario is notably more challenging due to item overlap and the need for 3D spatial analysis to determine both the position and orientation of food items. While 3D template matching can be used to locate and orient objects with clearly defined shapes in 3D space [9], applying this technique to food products presents difficulties due to their variable geometries. Consequently, accurate food recognition for diverse food items in 3D or RBP scenarios remains an unresolved and significant challenge.

1.2.3 Fundamental research

To handle food products effectively, a well-developed handling strategy is essential. For example, the grasping force must be carefully calibrated: it should be gentle enough to prevent damage to the food item, yet strong enough to complete a pick-and-place task without risk of slippage. Additionally, grasping velocity plays a crucial role, as it must account for the viscosity of the food product and the potential for impact-related damage during handling.

To the best of the authors' knowledge, limited research has focused on developing optimal handling strategies specifically for food products. In practical applications, these strategies are typically determined through trial-and-error experimentation due to the scarcity of foundational studies and detailed information on the "engineering" properties of food items. Such properties include size, shape, weight, softness, surface condition, friction coefficient, viscoelasticity, rheology, fragility, and susceptibility to bruising, among others.

While specialized machines and devices are available to measure these properties in fields such as food science, nutrition, and mastication, data specific to robotic handling purposes are sparse. However, these data are crucial for designing robotic end-effectors and refining grasping strategies. Furthermore, handling strategies vary based on both the robotic end-effector and the target food item. To improve the versatility of handling approaches, systematic categorizations of robotic end-effectors and food products based on their characteristics are essential; however, such research has been infrequent to date.

1.3 Book Structure

This book is dedicated to addressing the aforementioned challenges and presenting recent research conducted by the authors. Recognizing the primary bottleneck as the limited availability of robotic end-effectors, the book focuses on the development of robotic end-effectors for automation across several sectors: the food industry (Chapter 2), agriculture (Chapter 3), and kitchen and restaurant environments (Chapter 4). An overview of food recognition techniques for robotic handling is introduced, along with select examples (Chapter 5).

In terms of foundational research, mathematical modeling and the measurement of food products' physical properties are discussed in Chapter 6. Chapter 7 introduces a method for classifying food products based on handling-related physical properties and includes descriptions of the fabrication of food samples and handling tests conducted on these samples. A classification system for robotic hands designed for food manipulation is detailed in Chapter 8. Building on the classifications of both food products and robotic hands, this chapter also presents an approach for evaluating the performance of various robotic hands.

Additionally, Chapter 9 provides a back-casting case study to illustrate the process of designing a robotic hand specifically for the automatic packaging of breadcrumb-coated oysters – a system that was successfully implemented in a food production line. Finally, Chapter 10 suggests several areas for future research and development.

2

Robotic Hands for Food Industry

This chapter introduces various robotic hands developed for manipulation within the food industry, including bending-finger hands, string-based hands, circular membrane hands, and needle hands. Each of these robotic hands is designed to handle specific food items. This chapter provides detailed descriptions of the target food items, design and analysis of each robotic hand, materials used, fabrication processes, and experimental evaluations.

2.1 Bending Finger Hand

Bendable finger hands consist of multiple bendable soft fingers driven by pneumatic or fluidic pressure [10, 11].

2.1.1 3D printable bending finger hand

In Japan, the popularity of box lunches (bento) is high due to their convenience and variety. Each day, several million bento boxes are produced and consumed across the country. However, the packing of these lunches is still predominantly performed manually, owing to the fragility, diversity, and high deformability of food items [12]. To reduce labor costs, the food industry has a strong demand for automated lunch box packing systems.

A typical lunch box (Figure 2.1a) comprises multiple small dishes, each placed in a waterproof cup container. These containers (Figure 2.1b) are highly deformable and feature a frustum shape with a small circular base. Handling such containers filled with food presents significant challenges due to the irregular shapes and deformability of contents. Traditional rigid grippers and vacuum packing systems, which are commonly used in the food industry, struggle with this task; rigid grippers risk damaging the food, while vacuum systems require a flat surface to achieve effective suction. Therefore, new grasping strategies are needed to better address these handling requirements.

When we pick up a cup container filled with peanuts, we typically use all five fingers (Figure 2.1c), allowing for a stable grasp with minimal effort. Using three fingers (Figure 2.1d) can also lift the container, though it requires

DOI: 10.1201/9781003372431-2

(c) (d) (e)

FIGURE 2.1
A boxed lunch (a) with paper containers (b) and human hand grasping with five (c), three (d), and two fingers (e), respectively. Reproduced with permission [13]. Copyright 2016, IEEE.

more effort and results in greater deformation of the container compared to a five-finger grasp. Based on experience, picking up the container with only two fingers (Figure 2.1e) is often challenging; even if successful, the significant deformation caused may be unacceptable. In [13], a gripper design with three modular fingers was developed, enabling easy assembly and replacement.

2.1.1.1 Soft finger design

With a focus on easy fabrication using a multi-material 3D printer, three different finger designs were proposed. Design 1 (Figure 2.2a) features 11 smaller chambers and a larger chamber at the fingertip. The smaller chambers have a wall thickness of 1 mm, while the larger chamber has a wall thickness of 3 mm, making the fingertip stiffer than the rest of the finger to mimic the function

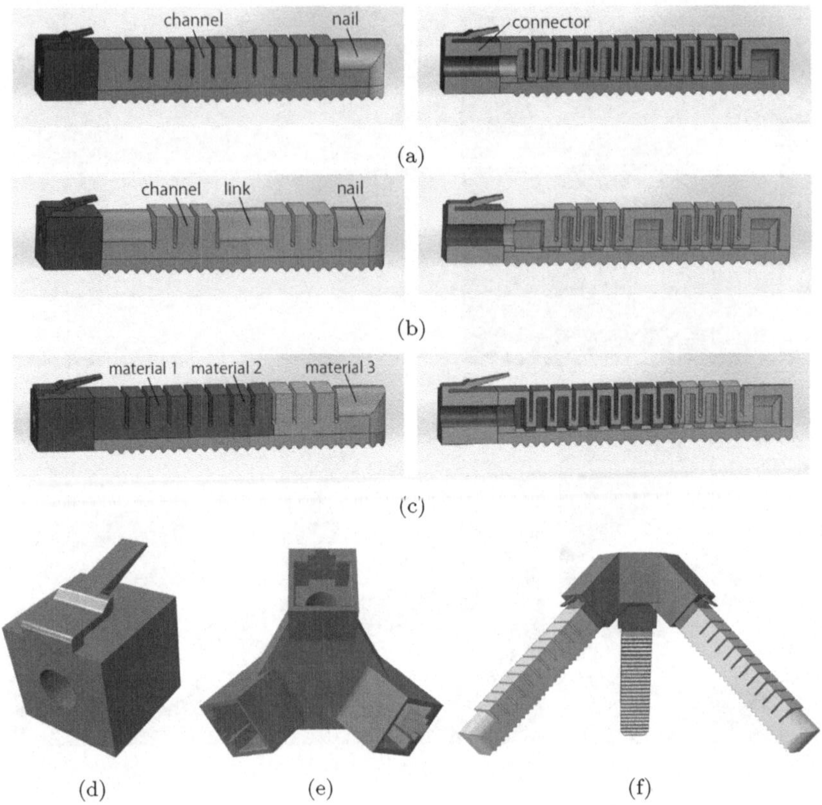

FIGURE 2.2
Different soft finger designs (a–c) and gripper components of connector (d), base (e) and assembly (f). Reproduced with permission [13]. Copyright 2016, IEEE.

of a human nail. Design 1 is intended to produce a uniform curvature along the finger's length.

To replicate the structure of a human finger with distal joints, Design 2 (Figure 2.2b) was developed. This design includes two links and six small chambers, imitating two human distal joints. Additionally, to explore variable curvature along the finger, Design 3 (Figure 2.2c) was created. This finger design consists of three regions with distinct material properties, where Material 1 is the hardest and Material 2 is the softest. To enhance grasping stability and mimic human fingerprints, a rippled texture was added to the bottom surface.

For ease of assembly, a connector (Figure 2.2d) and a base (Figure 2.2e) were designed. The connector can be 3D printed together with the soft finger, featuring a 6 mm-diameter hole for inserting an air hose. A snap-lock mechanism, with male (on the connector) and female (on the base) components, was implemented to allow easy attachment of the connector to the base. The base includes three female interfaces arranged in a circular pattern with equal spacing (Figure 2.2e). The fully assembled gripper is shown in Figure 2.2f, with the fingers positioned at a 45° vertical angle to form the initial gesture.

2.1.1.2 Finite element (FE) simulations of the soft fingers

Since it is difficult to analytically determine the dynamics of soft robots, finite element (FE) analysis is often used to investigate their behavior before fabrication and experimentation. In [13], the commercial FE software Abaqus (Simulia, Dassault Systèmes) was employed to analyze the deformation of soft fingers. The Young's moduli for Materials 1, 2, and 3 were set to 3.65 MPa, 1.55 MPa, and 2.25 MPa, respectively, based on he Shore hardness values provided in the material datasheet.

Figure 2.3 presents the simulation results of a single finger subjected to gravity and varying air pressures. The unloaded finger was initially positioned horizontally, and loading conditions were applied in two steps. In the first step, gravity was applied, and the resulting deformation was simulated, as shown at the top of Figures 2.3a–c. In the second step, air pressure was applied, and

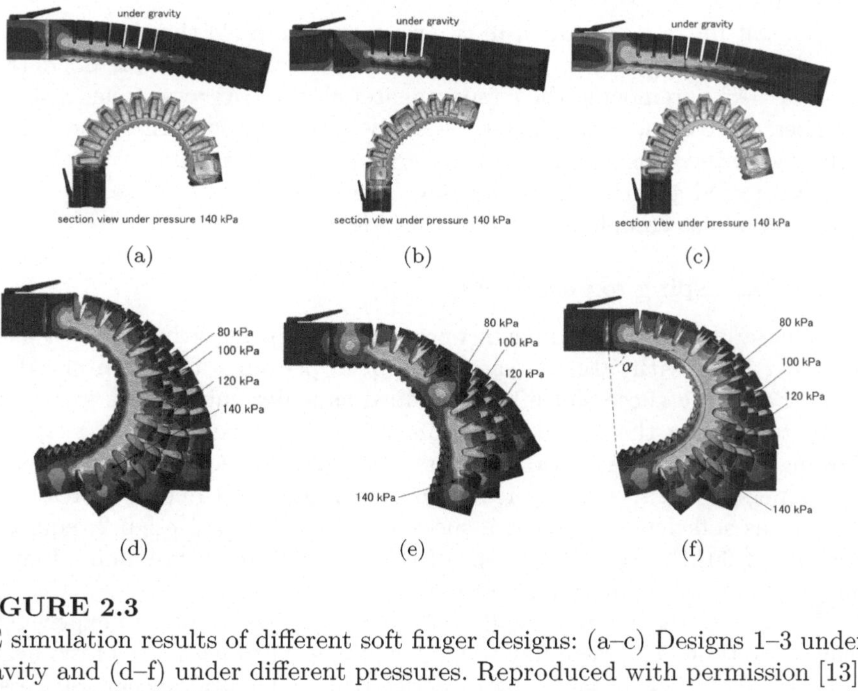

FIGURE 2.3
FE simulation results of different soft finger designs: (a–c) Designs 1–3 under gravity and (d–f) under different pressures. Reproduced with permission [13]. Copyright 2016, IEEE.

the simulation results are provided in Figures 2.3d–f. The stress distributions within the chambers are displayed at the bottom of Figures 2.3a–c.

The simulation results suggest that Design 2 exhibits higher stiffness, with less bending under both gravity and applied pressure, compared to other designs. This is due to Design 2 having fewer chambers than the other two designs. Design 3 shows greater bending under gravity, as it contains a softer region (Material 2) in the middle section. When pressure is applied, Design 1 generates the largest bending among the three designs. Compared to Design 1, Design 3 demonstrates reduced bending due to a stiffer region (Material 3) positioned near the base.

Additionally, Design 1 produces a uniform curvature, Design 2 bends primarily at the joint regions, and Design No. 3 displays non-uniform curvature along its length. The maximum stress levels are similar between Designs 1 and 2, but Design 3 exhibits significantly higher stress, particularly in the middle region where the softer material is located (Figure 2.3c).

2.1.1.3 Fabrication of the soft finger and gripper

The grippers were fabricated using a multi-material 3D printer (Objet260 Connex, Stratasys). This printer is capable of simultaneously printing 14 digital materials with varying mechanical properties by combining a soft, rubber-like material (TangoBlack+) with a hard, polypropylene-like material (VeroWhite).

The soft finger was printed in two separate parts: (1) the chambers along with the connector, and (2) the bottom cover, which seals the chambers (Figure 2.4a). After removing the support material, the cover was glued onto the chambers to complete the finger assembly (Figure 2.4c). The base (Figure 2.4b) was printed using the hard material (VeroWhite). Once the air hoses were connected and the three fingers were inserted into the base, the assembled gripper was completed, as shown in Figure 2.4d.

2.1.1.4 Grasping experiments

Grasping and lifting experiments were conducted using grippers on a paper container filled with 50 g of peanuts. Applied pressures ranged from 20 kPa to 60 kPa, and a successful lift was defined as maintaining a 10 s suspension without dropping the container. Figure 2.5 presents experimental snapshots showing the initial state prior to pressurization and successful lifts achieved by grippers with varying finger designs. With Designs 1 and 3, a pressure of 40 kPa was sufficient to achieve a successful lift, whereas Design 2 required a pressure of 60 kPa. Notably, gripper Design 3 exhibited a pronounced curvature, aligning with simulation results.

To evaluate the grasping ability of the gripper, we conducted experiments using various objects, including a small container filled with 50 g of red beans, a 36 g piece of fried chicken, a 64 g wooden ball, and a 48 g raw egg. Successful lifts achieved using Design 3 are shown in Figure 2.6. Lifting the container

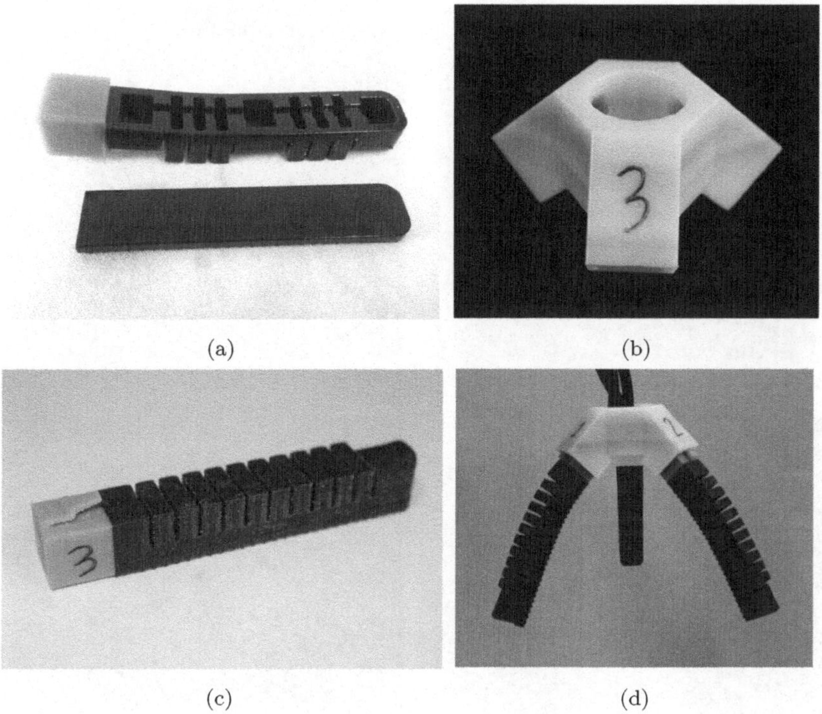

FIGURE 2.4
Fabrication and gripper assembly: (a) two printed finger components, (b) the printed base, (c) the completed soft finger by gluing the two components together, and (d) the assembled soft gripper. Reproduced with permission [13]. Copyright 2016, IEEE.

filled with beans proved more challenging, requiring a higher pressure (50 kPa) compared to lifting the container with peanuts. This difference is attributed to the smaller size of the bean container, necessitating greater bending to achieve comparable grasping and lifting force. A similar trend was observed when comparing the grasping and lifting of the wooden ball with those of the raw egg: the smaller egg required more effort to lift than the larger wooden ball. Additionally, it was generally easier to grasp and lift rigid objects, such as the wooden ball and the egg, compared to softer objects, as deformation in softer objects occurred during lifting.

2.1.2 Bending finger hand equipped with suction pads

A recent review [14] categorizes soft robotic grippers into three groups based on their grasping principles: (1) gripping by actuation, (2) gripping by controlled stiffness, and (3) gripping by controlled adhesion. Most grippers in the

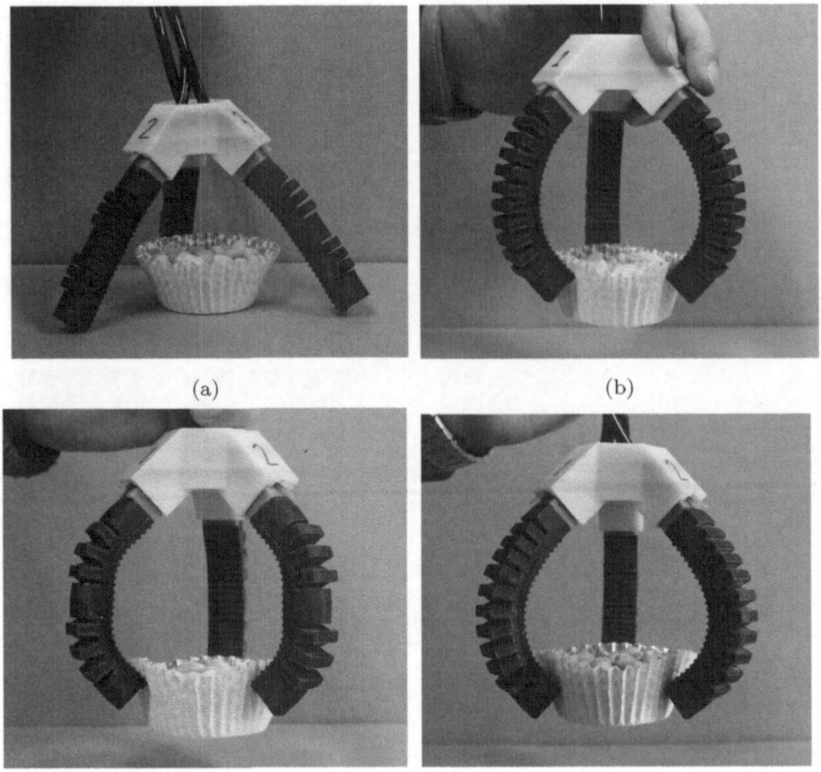

(a) (b) (c) (d)

FIGURE 2.5
Grasping experiments on a paper container filled with peanuts: (a) initial state, (b) grasping with finger Design 1, (c) grasping with finger Design 2, and (d) grasping with finger Design 3. Reproduced with permission [13]. Copyright 2016, IEEE.

first two categories rely on static friction to lift objects, which typically requires a large contact area to generate sufficient friction. Consequently, these grippers often struggle to grasp thin objects effectively. In contrast, grippers in the third category excel at picking up thin objects through surface adhesion, including grippers that use suction [15, 16], electrostatic adhesion [17], and gecko-inspired adhesion [18, 19]. However, these adhesion-based grippers face limitations when handling deformable objects or objects with rough or wet surfaces. To integrate the strengths of both approaches, a dual-mode soft gripper that incorporates suction pads on the tips of pneumatic soft fingers was proposed in [20]. By combining the advantages of multi-fingered grasping with suction capabilities, the proposed gripper is designed to handle a wide variety of objects.

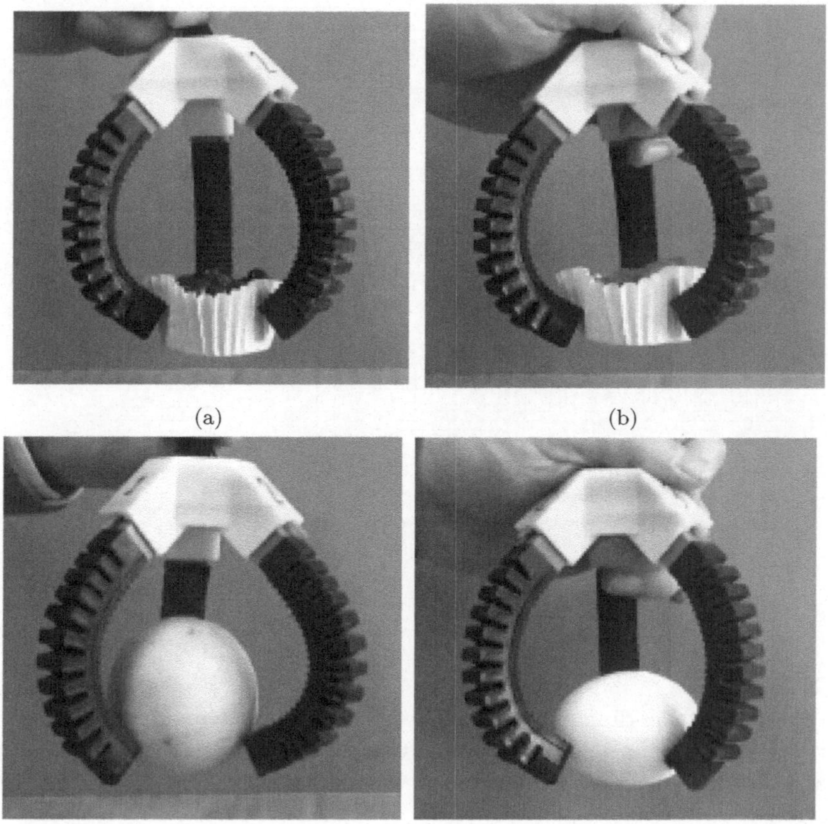

(a) (b) (c) (d)

FIGURE 2.6
Experimental snapshots of grasping various objects: (a) a paper container filled with red beans, (b) a fried chicken in a paper container, (c) a wooden ball, and (d) a raw egg. Reproduced with permission [13]. Copyright 2016, IEEE.

2.1.2.1 Soft finger design

The design of the soft finger is illustrated in Figure 2.7. Excluding the suction pad, the finger measures 95 mm × 20 mm × 18 mm (L × W × H), making it slightly smaller than a human finger. The finger consists of four main components (Figure 2.7a): the chamber, air paths, cover, and suction pad. As shown in Figure 2.7c, the chamber comprises 13 air chambers, each separated by 1.5 mm-thick walls. The chamber cavity has a thickness of 2 mm, with an interval of 1.5 mm between adjacent chambers. The air paths, shown in Figure 2.7d, include two main pathways with a cross-sectional dimension of 2 mm × 2 mm, designed to connect all chambers for inflation and deflation.

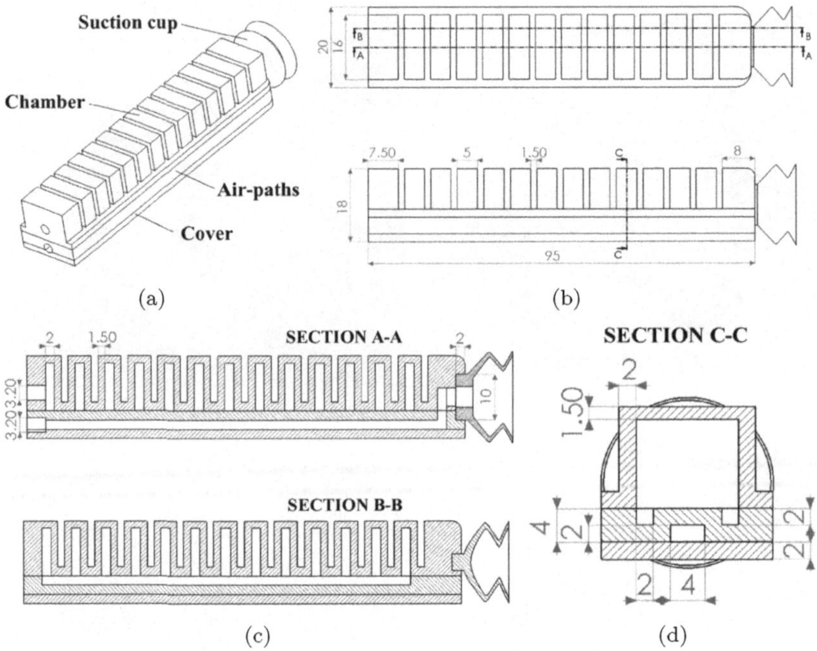

FIGURE 2.7
CAD design of bending finger hand with suction pad in: (a) the isometric view, (b) the top and side views, (c) the section views A-A and B-B, and (d) the section view C-C. Reproduced with permission [20]. Copyright 2020, Elsevier.

Additionally, an air path leading to the suction pad has a cross-sectional dimension of 4 mm × 2 mm. The cover, which seals the air paths, has a thickness of 2 mm. Commercially available suction pads with a single bellow are selected for their performance, and a circular indentation (10 mm diameter and 2 mm depth) is incorporated at the fingertip to secure the suction pad.

2.1.2.2 Soft finger fabrication and gripper prototype

The soft finger was fabricated using a traditional casting process. Casting molds (Figure 2.8a) were 3D-printed with a Zortrax M200 printer (Zortrax, Olsztyn). The tools and devices used in the casting process are shown in Figure 2.8b. Fabrication was divided into several sequential steps, as illustrated in Figure 2.8c. First, the mixed liquid rubber was degassed for two minutes at a vacuum pressure of approximately −85 kPa. Following degassing, the liquid rubber was poured into the chamber female mold (step A), the air-paths female mold (step B), and the cover mold (step C). These molds, filled with liquid rubber, were then placed in the vacuum chamber for a second round of degassing. Subsequently, the chamber male mold and air-paths male

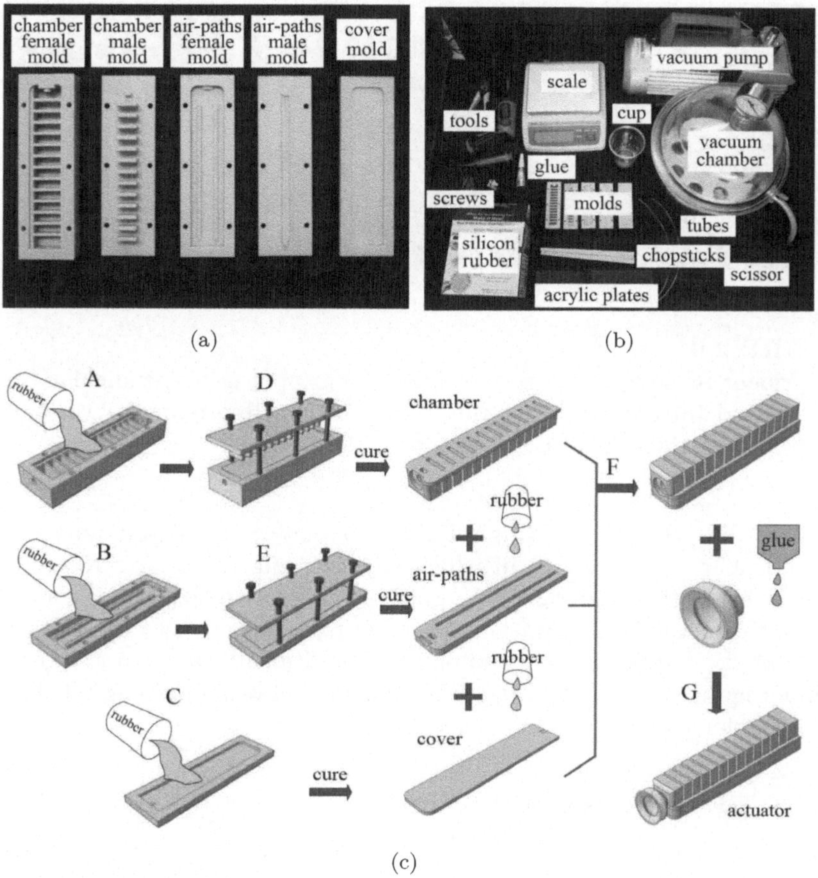

FIGURE 2.8
Soft finger fabrication: (a) 3D printed molds, (b) devices and components used for casting, and (c) fabrication process. Reproduced with permission [20]. Copyright 2020, Elsevier.

mold were inserted into their respective female molds and secured with screws (steps D and E). The cover mold was sealed with an acrylic plate. Once cured, the chamber, air-paths, and cover components were carefully demolded. The three parts were then bonded together in layers using the same liquid rubber, completing the soft finger body (step F). In the final step (step G), a commercial suction pad (FGA 20 SI, Schmalz, Germany) was affixed to the circular indentation at the fingertip with instant adhesive (Cemedine Co., Ltd., Tokyo).

A gripper base (Figure 2.9a) was fabricated by 3D printing with an Objet350 printer (Stratasys, MN, United States) using rigid VeroWhite material. The base includes two series of positioning holes, enabling both perpendicular

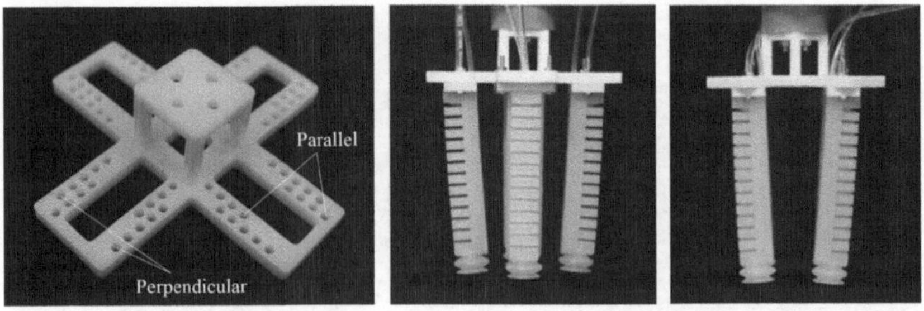

(a) (b) (c)

FIGURE 2.9
Soft gripper prototype: (a) gripper base, (b) gripper in perpendicular configuration, and (c) gripper in parallel configuration. Reproduced with permission [20]. Copyright 2020, Elsevier.

(Figure 2.9b) and parallel (Figure 2.9c) configurations of soft fingers. The perpendicular configuration, in which adjacent fingers are oriented at right angles to each other, is designed for grasping circular and square objects. Conversely, the parallel configuration, where adjacent fingers are aligned parallel, is suitable for handling elongated objects. Additionally, a 5 mm gap between neighboring holes allows for adjustable spacing between opposing soft fingers, enabling adaptation to target objects of varying sizes.

2.1.2.3 FE simulation of the soft finger

Abaqus (SIMULIA, Dassault System, MA) was used to simulate the performance of the soft finger. Two materials, Dragon Skin 10 (DS10, Smooth-On Inc., PA) and Dragon Skin 30 (DS30), were employed to model the finger. Based on tensile tests and model fitting, the first-order Ogden model was selected to characterize the nonlinear behavior of the materials. The identified material parameters are $\mu_1 = 0.2640 \pm 0.0078\,\text{MPa}$, $\alpha_1 = 3.0158 \pm 0.0453$, and $D_1 = 0\,\text{MPa}^{-1}$. The finite element model of the finger was meshed using 10-node quadratic tetrahedral hybrid elements (C3D10H) with a global mesh seed size of 1 mm. Gravity was included in the simulations, with material densities set to 1,070 kg/m^3 for DS10 and 1,080 kg/m^3 for DS30, as specified in the material datasheets. Pressure loads (inflation, deflation, and suction) were applied to the relevant surfaces, and contact interactions were defined for surfaces in mutual contact. Standard implicit simulations were conducted with geometrical nonlinearity enabled to account for large bending deformations.

Simulation results of finger inflation are presented in Figures 2.10a and c. The soft finger made from DS10 material achieves an approximate 90° bend (angle α in Figure 2.10a) at a pressure of 24 kPa, with the bending curvature increasing progressively toward the suction pad end. In contrast, the finger

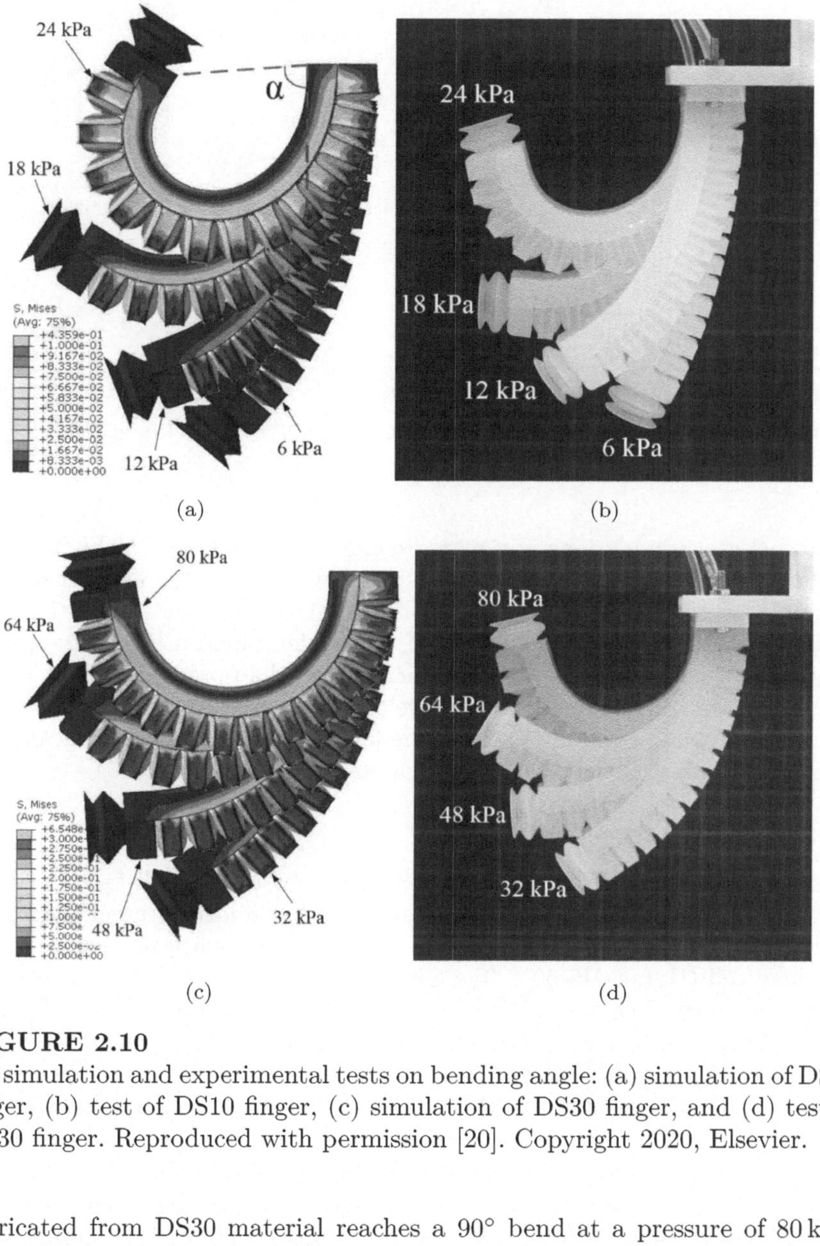

FIGURE 2.10
FE simulation and experimental tests on bending angle: (a) simulation of DS10 finger, (b) test of DS10 finger, (c) simulation of DS30 finger, and (d) test of DS30 finger. Reproduced with permission [20]. Copyright 2020, Elsevier.

fabricated from DS30 material reaches a 90° bend at a pressure of 80 kPa, displaying a nearly constant bending curvature along its length (Figure 2.10c).

For experimental validation, the same pressures applied in the simulations were used to inflate soft fingers. Bending behaviors were recorded using a digital camera, with examples shown in Figures 2.10b and d. Four fingers fabricated from DS10 and DS30 materials, respectively, were tested to assess individual variability among the fingers. Quantitative comparisons of bending

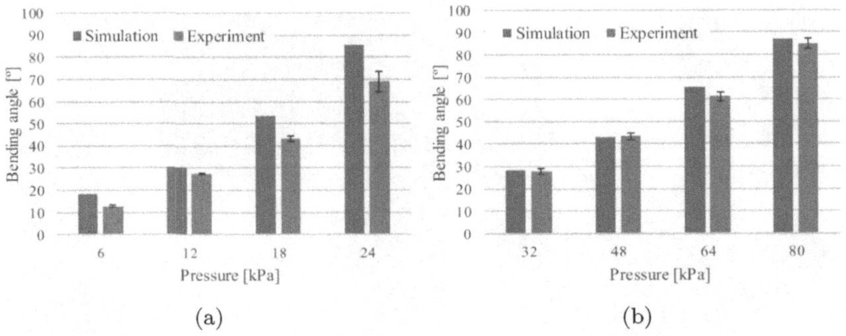

FIGURE 2.11
Bending angle comparisons of FE simulation and experimental tests: (a) DS10 finger and (b) DS30 finger. Reproduced with permission [20]. Copyright 2020, Elsevier.

angles between simulations and experiments are summarized in Figure 2.11. For fingers made with DS10, discrepancies between simulated and experimental results were relatively large, suggesting that the material properties used in simulation did not fully capture the actual deformation behavior; specifically, the actual finger exhibited greater stiffness than predicted. In contrast, simulations and experiments aligned well for fingers made with DS30. Additionally, individual variability among the fingers was minimal, as indicated by the small standard deviations in bending angles.

2.1.2.4 Lifting force tests

To demonstrate the lifting capability at different opening distances, lifting tests with force measurements were conducted. A cylindrical target (66 mm in diameter and 50 mm in length), 3D-printed using ABS material, was mounted on a load cell (USL06-H5-50N, Tec Gihan, Kyoto) (Figure 2.12a) to measure the lifting force. The soft gripper was attached to a robot arm (VP-6242, Denso, Tokyo), which was programmed to execute a simple picking motion. In the initial position (Figure 2.12a), the gripper was deflated to achieve a large opening. The gripper was then translated downward to the grasping position (Figure 2.12b), where it was inflated to grasp the target. Subsequently, the gripper was translated upward at a speed of 390 mm/s (Figure 2.12c). Lifting tests were performed at various pressures and opening distances in the perpendicular configuration. The deflated gripper generated opening ranges (diameters it could accommodate) of approximately 128.6 mm–176.9 mm for the DS10-based gripper, and 139.5 mm–193.6 mm for the DS30-based gripper. The DS10 gripper exhibited a smaller opening range due to the effect of gravity, which partially counteracted the deflation owing to the material's softness. Five trials were conducted for each input pressure.

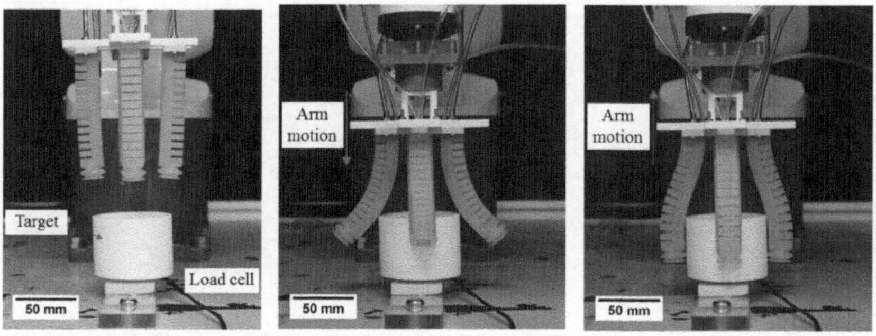

(a) (b) (c)

FIGURE 2.12
Lifting force test: (a) experimental setup and initial position, (b) the grasping position, and (c) the lifting state. Reproduced with permission [20]. Copyright 2020, Elsevier.

To quantitatively compare the differences in lifting force, we recorded the maximum lifting force (absolute value) for each trial, along with their average and standard deviation. Figure 2.13 presents a comparison of the maximum lifting force for different opening distances and materials. It is evident that a smaller initial opening results in a larger lifting force, with this trend being more pronounced in the gripper fabricated from the harder material (Figure 2.13b). This is because a smaller initial opening creates a tighter contact with the target, leading to more stable grasping. Furthermore, a linear relationship between lifting force and air pressure is more clearly observed in the harder gripper, and the standard deviations are smaller in Figure 2.13b compared to Figure 2.13a. Consequently, the gripper made from the harder material

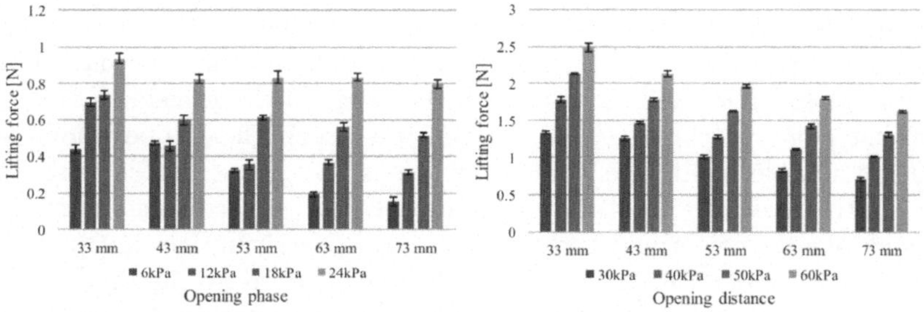

(a) (b)

FIGURE 2.13
Lifting force comparisons for different opening distances: (a) DS10 gripper and (b) DS30 gripper. Reproduced with permission [20]. Copyright 2020, Elsevier.

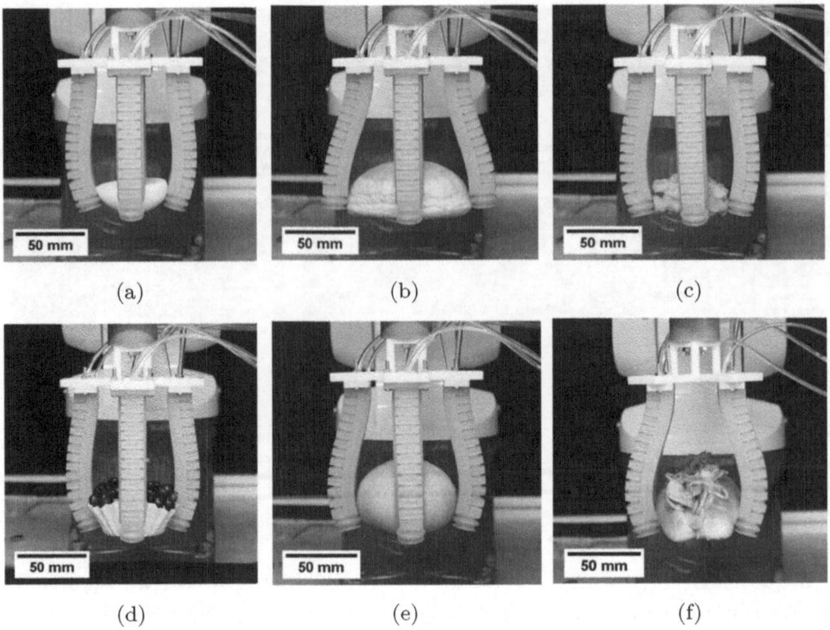

FIGURE 2.14
Experimental snapshots of grasping tests: (a) half a boiled egg, (b) bread, (c) fried chicken, (d) red beans in a paper container, (e) an orange, and (f) a hot dog. Reproduced with permission [20]. Copyright 2020, Elsevier.

demonstrates more consistent and robust performance than the one made from the softer material.

2.1.2.5 Grasping tests

Grasping tests on various food materials were conducted using two grippers fabricated from DS10 and DS30. The perpendicular configuration was employed to grasp all targets, except for the hot dog. Due to the elongated shape of the hot dog, the parallel configuration was used. The gripper made with DS10 was unable to lift the hot dog, whereas the gripper made with DS30 successfully lifted it at a pressure of 70 kPa. For all other targets, both grippers successfully performed the grasping tests, requiring varying pressures. Notably, the gripper with DS10 required lower pressure to lift the same target, suggesting that the gripper with softer fingers is more energy-efficient. Figure 2.14 shows snapshots of the grasping tests using the DS30 gripper.

2.1.2.6 Suction tests

Suction tests were conducted on various targets using the grippers fabricated from both DS10 and DS30 materials. The sauce and kernel corn were packaged

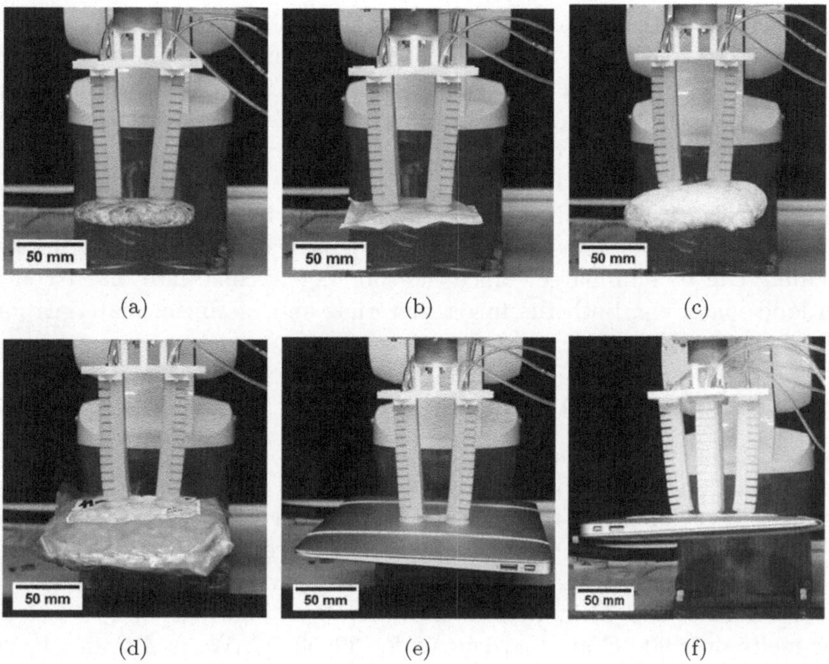

FIGURE 2.15
Experimental snapshots of suction tests: (a) a frozen hamburger, (b) a bag of sauce, (c) wrapped rice, (d) a bag of kernel corn, (e) a MacBook Air using the DS30 gripper in parallel configuration, and (f) the same MacBook Air using the DS10 gripper in perpendicular configuration. Reproduced with permission [20]. Copyright 2020, Elsevier.

in plastic bags, while the cooked rice was wrapped in cling film. A vacuum ejector (VUE07-44A, PISCO, Japan) was used to generate negative pressure for each finger. Pressures of approximately −60 kPa and −90 kPa were applied for the gripper made from DS10 and DS30, respectively, during the suction tests. Both grippers successfully lifted all targets stably in both finger configurations. Some snapshots of the suction tests are shown in Figure 2.15. These thin and heavy targets are typically difficult to grasp, but by leveraging the suction pads, successful handling was achieved. As the suction pads on each of the four fingers are independently vacuumed, the success rate is maximized. The suction pads proved effective not only on smooth surfaces, such as the MacBook Air, but also on a frozen hamburger and plastic bags.

2.1.3 Gluing-free fabrication of bendable fingers

This section introduces a glue-free fabrication method for bendable fingers. Since soft fingers come into contact with food materials during food manipulation, it is essential that the materials used for these fingers are food-compatible. In traditional molding-and-gluing techniques, commonly applied in the fabrication of PneuNet actuator-based soft fingers, multiple parts are fabricated through molding and then glued together. However, the adhesives used in these processes are often not food-compatible, making them unsuitable for such applications. In contrast, 3D printing enables the use of sacrificial materials to fabricate complex-shaped structures without the need for gluing. By incorporating sacrificial materials into mold-based fabrication, the need for gluing can be eliminated. Since the soft fingers come into direct contact with food materials, both the finger materials and sacrificial materials must be food-compatible.

2.1.3.1 Gluing-free fabrication approach

Figure 2.16 shows a bendable finger hand for topping of strawberries on a cake base [21]. This hand is composed of a pair of bendable soft fingers. Soft fingers are cast in silicone rubber using gelatin as a sacrificial material. There are several candidates of food-compatible sacrifice materials. For example, gelatin and agar come under this category as they harden when cooled and melt when heated. Gelatin melts over 50–60 °C and hardens at 20 °C or less. Agar melts over 90 °C and hardens under 30–50 °C. We find that gelatin is easier to use than agar. Thus, we applied gelatin as a sacrifice material. Both silicone rubber and gelatin are food compatible. Here glue-free fabrication is applied to exclude non-food compatible materials.

Figure 2.17 shows glue-less fabrication process of soft fingers. Finger material was liquid silicone rubber Dragon Skin 30 and sacrifice material was gelatin. First, a mold for the chamber side of a PneuNet actuator was fabricated. Liquid silicone was poured into the mold, and the silicone in the

FIGURE 2.16
Bendable finger hand for cake topping. Reproduced with permission [21]. Copyright 2024, IEEE.

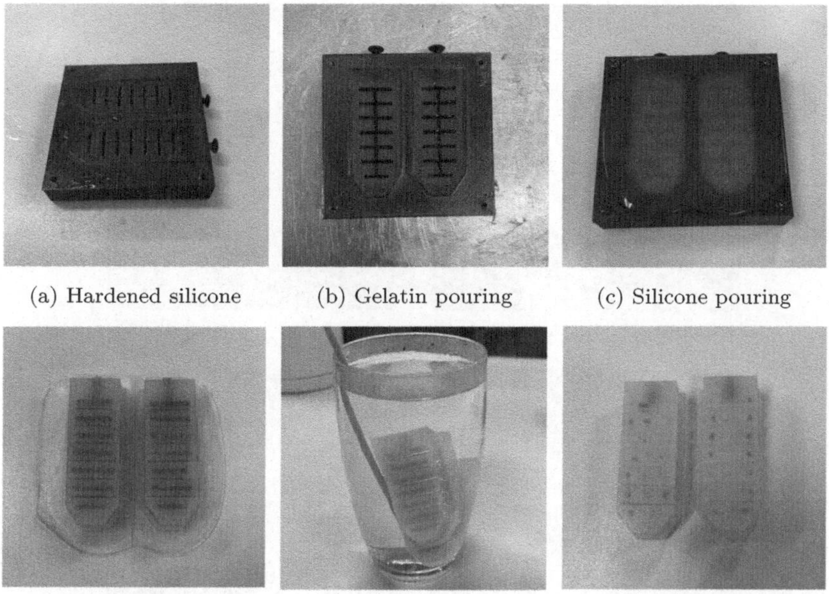

(a) Hardened silicone (b) Gelatin pouring (c) Silicone pouring

(d) Demolding (e) Dissolving gelatin (f) Fabricated finger

FIGURE 2.17
Fabrication of gluing-free soft fingers. Reproduced with permission [21]. Copyright 2024, IEEE.

mold was hardened. Figure 2.17a shows the mold filled with hardened silicone. Cavities in the hardened silicone correspond to chambers. Second, gelatin was liquefied in hot water. The liquefied gelatin was mixed with food coloring and then poured into the cavities in the hardened silicone. Excess gelatin was wiped off. The poured gelatin was solidified by chilling it in a refrigerator for 30 minutes. Figure 2.17b shows the hardened gelatin in cavities. Third, liquid silicone was poured over the hardened silicone and the hardened gelatin. This poured liquid silicone worked as a cover. The poured liquid silicone hardened in room temperature, as shown in Figure 2.17c. Then, molded fingers were detached from the mold. Figure 2.17d shows a detached finger with solidified gelatin. Excess silicone parts were cut off. Detached fingers were then immersed in hot water to liquefy the solidified gelatin in fingers, as shown in Figure 2.17e. The liquefied gelatin was removed from fingers. Finally, we have a soft finger, as shown in Figure 2.17f.

2.1.3.2 Cake decoration test

Figure 2.18 shows results of topping. Strawberry topping was performed on cake bases with two coating styles, that is, on a nappe base (Figure 2.18a)

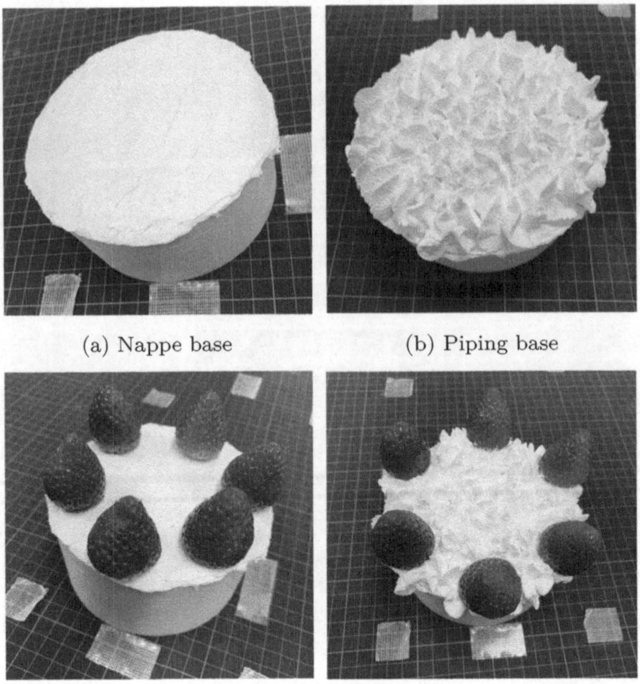

(a) Nappe base (b) Piping base

(c) Topping on nappe base (d) Topping on piping base

FIGURE 2.18
Topping results. Reproduced with permission [21]. Copyright 2024, IEEE.

and a piping base (Figure 2.18b). In nappe coating, cream is coated over a cake base uniformly, resulting in a flat top surface. In contrast, in piping style, cream is squeezed out from a piping bag and is coated over a cake base, resulting in an uneven top surface. The bendable finger hand was able to grasp real strawberries and place them to their goal positions on a nappe base one by one successfully. The final state is shown in Figure 2.18c. The bendable finger hand was able to grasp real strawberries and place them to their goal positions on a piping base one by one successfully. Figure 2.18d shows the final state of strawberry topping on the piping base. No strawberries fell, but several strawberries slanted slightly after placing. This is due to the unevenness of the piping base. Due to softness of coated cream, strawberries remained on the cake base. In conclusion, the topping on a nappe base and a piping base was successful in spite of several strawberries slanted slightly after placing on a piping base.

2.2 Variations of Bending Finger Hands

This section presents various designs for bending finger hands. By carefully designing the structure and shape of the bending fingers, we can expand the range of food items that can be grasped. Specifically, we introduce prestressed hands for handling thin food items and wrapping hands for handling chopped food items.

2.2.1 Prestressed hand

The standard soft gripper with multiple straight fingers, as introduced in section 2.1.1, can grasp various objects, including a highly deformable paper container filled with peanuts. However, due to the initial configuration angle (Figure 2.5a), the contact area during grasping is limited, which affects the stability of the grasp (Figure 2.5b). To increase the contact area, a parallel configuration is often preferred, as demonstrated by [22–24]. However, the parallel configuration limits the initial opening space, meaning only objects that fit within this space can be grasped. Otherwise, a vacuum system is required to deflate the actuator before grasping, which creates a larger opening space [24]. This approach also necessitates a control system to switch between air inflation and deflation. To address this, a prestressed soft gripper was proposed in [11], which maintains the same contact area (Figure 2.19b) and creates a larger initial opening space (Figure 2.19a) compared to grippers with a parallel configuration. Furthermore, it only requires a single air system to inflate the soft actuators.

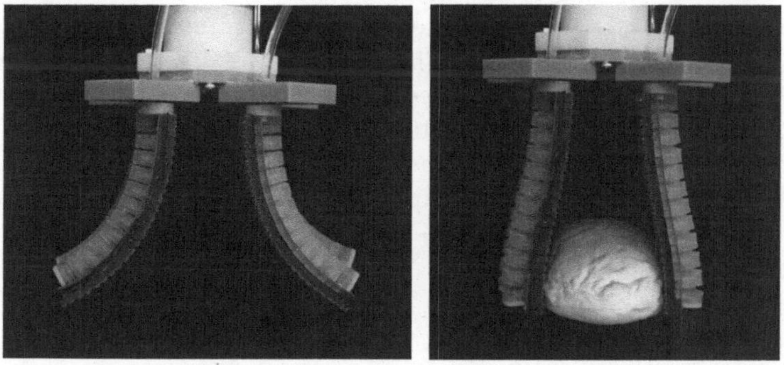

(a) (b)

FIGURE 2.19
Prestressed soft hand at its (a) initial and (b) grasping states. Reproduced with permission [11]. Copyright 2017, IEEE.

2.2.1.1 Concept

In nature, layered structures with varying physical properties (e.g., expansion and contraction) give rise to natural folding [25] or wrinkling [26] phenomena, such as those observed in drying leaves or grapes. These phenomena are driven by prestress generated at the interface between two layers. Prestress is commonly employed in concrete engineering to enhance the strength and fatigue resistance of concrete slabs [27]. In [11], prestress was induced by stretching the soft chamber part and gluing a non-stretched cover onto it. Upon releasing the tension, the actuator naturally curled toward the chamber side due to the induced prestress.

2.2.1.2 Curling simulation of prestressed finger

To estimate the curling deformation of the prestressed soft finger, finite element (FE) simulations were performed using Abaqus (Dassault System, MA). The simulation was conducted in two steps. In the first step, the stretching of the soft chamber part was simulated without considering the cover. After the stretching process, the deformed soft chamber part and the corresponding prestress distribution from the finite element model were transferred to the second step, where the cover part was attached to the soft chamber. A stress relaxation simulation was then carried out. Upon completion of the second step, the generated prestress was partially released, causing the curling deformation of the soft finger. The final states of the soft fingers for different stretching distances of 6 mm, 10 mm, and 14 mm are shown in Figure 2.20a.

2.2.1.3 Fabrication of prestressed finger

To fabricate the soft finger, the chamber and cover parts were 3D printed using an Objet260 3D printer (Stratasys, MN, United States) with a soft, rubber-like material (TangoPlus). The chamber was stretched using a linear stage, after which the cover was glued to the top of the soft chamber using a rubber-based adhesive (ThreeBond 1521B). The resulting curled fingers, with different stretching distances of 6 mm, 10 mm, and 14 mm, are shown in Figure 2.20b. A comparison between the experimental results (Figure 2.20) and the simulation data reveals good agreement between the simulated and fabricated soft fingers.

2.2.1.4 Grasping tests

Grasping tests were conducted on various food items, including paper containers filled with spaghetti, higiki, and ohitashi, commonly found in Japanese lunch boxes. For each target, ten pick-and-place trials were performed, and a successful lift was defined as a 10 s hold without dropping the item. Examples of successful trials are shown in Figure 2.21. Grasping and lifting food materials in paper containers were relatively stable, and no difficulty was encountered

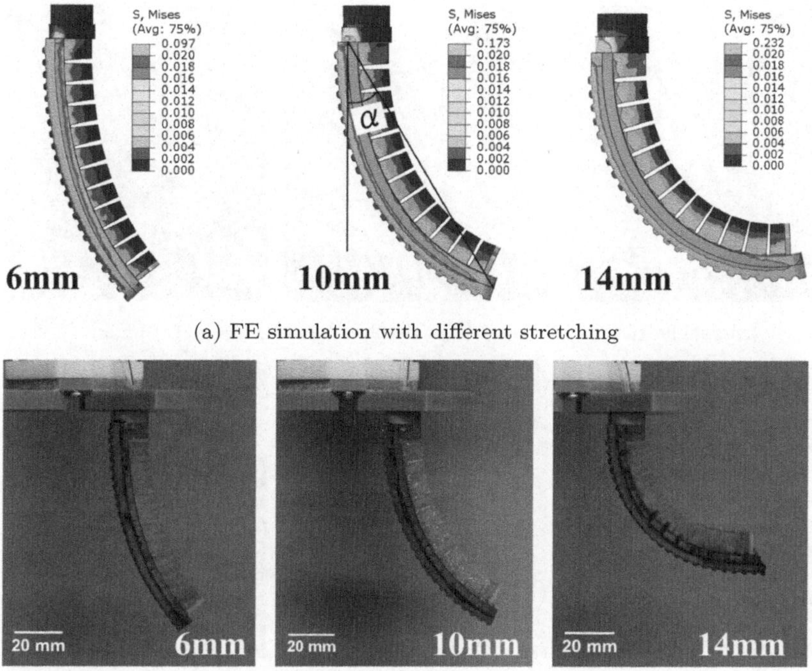

(b) Bending test results with different stretching

FIGURE 2.20
Results of (a) FE simulations and (b) bending tests with different stretching. Reproduced with permission [11]. Copyright 2017, IEEE.

when lifting a piece of omelet. However, the success rate for picking up targets with irregular shapes, such as fried chicken and salmon, was lower compared to other targets. Increasing the input pressure improved the success rate for lifting the fried chicken but did not significantly affect the performance when handling the salmon, due to its thin and irregular geometry.

2.2.2 Wrapping hand

Multi-fingered soft grippers are capable of handling a wide range of targets of different types and sizes, such as fried chicken, raw eggs, and baked salmon. However, due to the gaps between neighboring fingers, these grippers face challenges in directly grasping finely chopped and granular food materials, such as chopped green onions, corn, and salad. These types of food materials are commonly found in boxed meals, and the packaging process is still largely performed by human workers in food factories. To address this challenge, a wrapping hand capable of forming a closed cavity was proposed in [28].

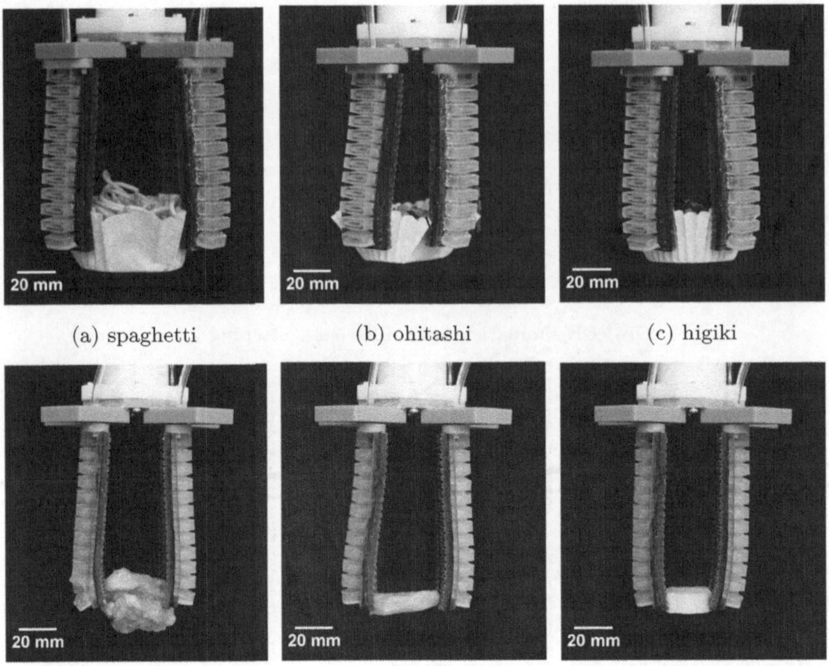

(a) spaghetti (b) ohitashi (c) higiki

(d) fried chicken (e) salmon (f) omelet

FIGURE 2.21
Grasping tests by prestressed soft hand on (a) spaghetti in a large rectangular container, (b) ohitashi in a small oval container, (c) higiki in a circular container, (d) fried chicken, (e) salmon, and (f) omelet. Reproduced with permission [11]. Copyright 2017, IEEE.

2.2.2.1 Wrapping hand design

As shown in Figure 2.22, the soft finger is designed with a triangular shape, having a tip angle of 90°. The overall size of the finger is determined based on the desired weight of the target to be grasped. Once the overall size is established, the volume of the cavity formed by the gripper can be approximated. The soft finger features five vertical and four horizontal air chambers, with a 90° tip angle to ensure complete closure when used with four fingers. Each air chamber has an external width of 3 mm, and the gap between adjacent chambers is set to 1 mm. The chamber wall thickness is designed as 1 mm for the preferable inflation direction and 2 mm for the non-preferable direction. The height of the horizontal chambers gradually decreases from 8 mm to 5 mm toward the finger tip. This design results in a thinner tip (5 mm in height), which facilitates easier insertion into food materials.

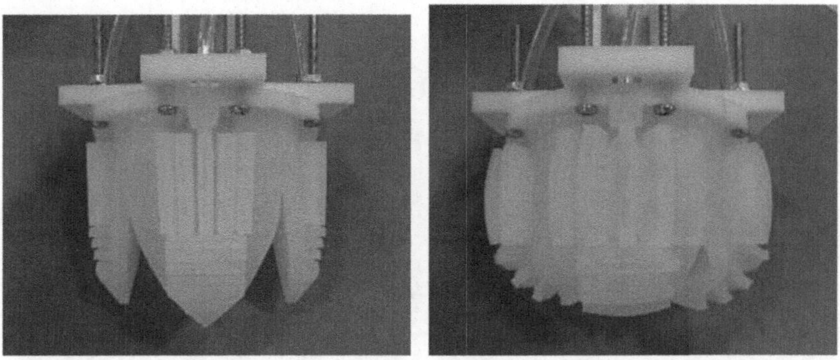

(a) (b)

FIGURE 2.22
Motion of wrapping hand at its (a) initial and (b) closed states. Reproduced with permission [28]. Copyright 2019, IEEE.

2.2.2.2 Grasping tests on corn and chopped onion

The grasping tests were first conducted on two food materials: chopped green onion and granular kernel corn. The materials were randomly scattered in a container, and the surface was manually flattened before testing. The surface height was input as a known parameter into the control program of the robotic manipulator. The wrapping hand was then controlled based on this height value. One key control parameter was the insertion depth, which indicates how far the distal part of the finger should be inserted below the material surface.

Figure 2.23 shows a series of experimental snapshots of the wrapping hand grasping kernel corn. After inserting the fingers into the corn, air pressure was applied to close the fingers, enabling the gripper to grasp the corn due to the small gap between neighboring fingers. The test results of the grasped weight are shown in Figure 2.24 for both corn and chopped onion, respectively. Data from three production line workers are also provided in Figure 2.24a for comparison.

It was found that the wrapping hand grasped more corn on average than the workers, with a lower standard deviation, indicating the stability of the wrapping hand. The results of grasping chopped onion, shown in Figure 2.24b, demonstrated that the wrapping hand grasped over 20% of the targeted weight, with a standard deviation of approximately 11.4%, which is acceptable for practical applications in production lines.

2.2.2.3 Improved hand for grasping salt

As shown in Figure 2.22b, the closed wrapping hand still leaves small gaps between neighboring fingers. This makes it difficult for the hand to grasp fine

(a) (b) (c) (d)

FIGURE 2.23
Snapshots of wrapping hand picking up corn. Reproduced with permission [28]. Copyright 2019, IEEE.

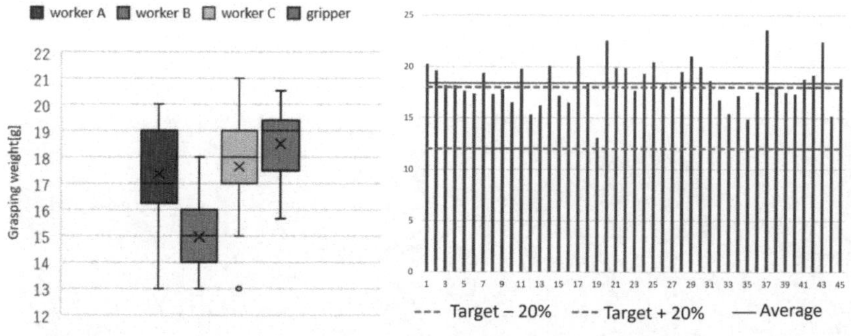

(a) (b)

FIGURE 2.24
Results of (a) human workers grasping corn and (b) wrapping hand grasping chopped onion.

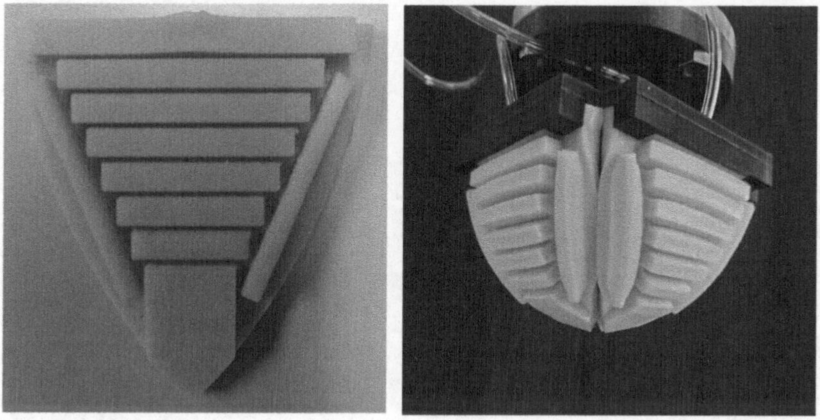

(a) (b)

FIGURE 2.25
Wrapping hand with oblique chambers: (a) soft finger and (b) hand closing with minimal gaps.

granular objects, such as salt. To address this issue, the wrapping hand was improved by adding oblique chambers on both sides of the finger, as shown in Figure 2.25a. Upon pressurization, the wrapping hand with oblique chambers minimizes the gaps between neighboring fingers, as shown in Figure 2.25b. The improved wrapping hand was then tested for grasping salt, with a series of experimental snapshots shown in Figure 2.26. The salt was successfully grasped without spilling.

FIGURE 2.26
Snapshots of wrapping hand with oblique chambers for picking up salt.

2.3 String-Based Hand

This section introduces string-based soft hands for food manipulation. When soft strings come in contact with a grasped object, the strings deform to prevent excess reaction forces. Also, soft strings deform according to the shape of the object. This increases the contacting area between the strings and the object to decrease the stress applied to the object. Thus, robotic hands using soft strings are effective to grasping delicate food items. In this section, we introduce binding hands and scooping-binding hands for food manipulation.

2.3.1 Binding hand

2.3.1.1 Concept

Binding hands apply elastic strings to grasp soft and deformable food materials [12]. Figure 2.27 shows a lunch box found in convenience stores and supermarkets in Japan. Each box has several food materials. Small-sized food materials such as beans and pickles are usually in cups made of plastic or waterproof papers (see the left center and top right of the figure). Currently, cups with food materials are manually packed into lunch boxes. Automatic packing is required to reduce cost and maintain sanitary. One challenge is to pick up such cups with food materials and place them into their desired locations.

In picking up an object, an external force must be applied to it such that the resultant lift force exceeds the gravitational force of the object. Usually, fingertips of a human or robotic hand are in contact with the object, with forces applied to the object through the contacting area between each fingertip

FIGURE 2.27
Lunch box with paper cups.

and the object. These applied forces can result in deformation of the object. This deformation can be reduced by increasing the area of contact, reducing the stress applied to the object by fingertips. Thus, rather than grasping a cup with food materials with rigid fingertips, one or multiple elastic strings surrounding a cup will be effective to grasping. The concept of *binding* was originally utilized to manipulate objects on the ground, using ropes operated by mobile robots [29]. This concept was applied to the grasping of cups with food materials.

2.3.1.2 Binding hand prototype

A binding hand was designed to pick up cups with food materials. This binding hand consists of multiple rigid fingers and an elastic string surrounding the fingers (Figure 2.28). One motor closes/opens the fingers, and another motor winds/relaxes the string synchronously to maintain the tension of the string. This finger set covers a square with a maximum size (Figure 2.29a) while occupying a square of minimum size (Figure 2.29b). The elastic string contacts a target object to be grasped instead of rigid fingers. This increases the contact area during grasping, contributing less deformation of the target object. Figure 2.30 demonstrates the process of binding hand picking a paper cup with beans. Paper cups are easily deformed. Additionally, beans in a cup easily change their relative location during manipulation. As a result, it is difficult to grasp and pick up a paper cup with beans. In this demonstration, a cup with beans

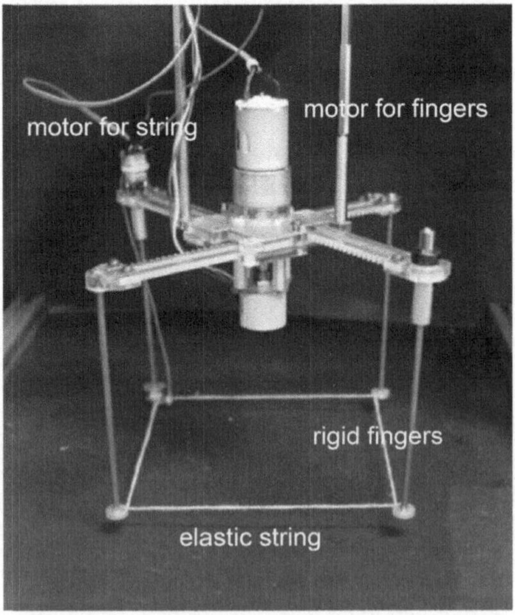

FIGURE 2.28
Binding hand. Reproduced with permission [12]. Copyright 2015, IEEE.

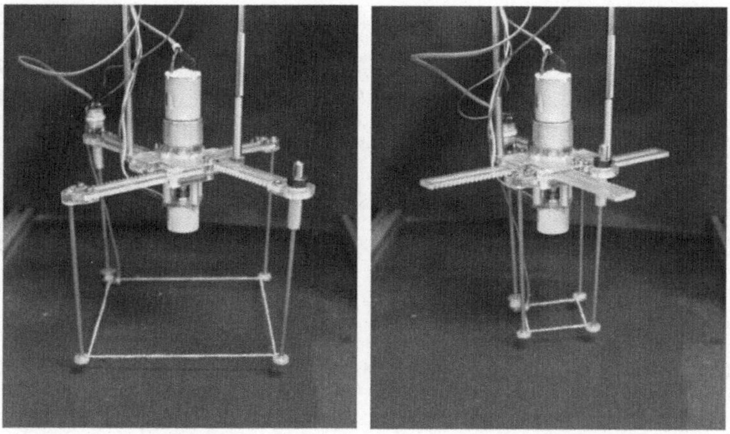

(a) Open (b) Close

FIGURE 2.29
Motion of binding hand. Opening/closing a set of rigid fingers with relaxing/winding an elastic string synchronously. Reproduced with permission [12]. Copyright 2015, IEEE.

is inside the region surrounded by an elastic string initially (Figure 2.30a). A set of rigid fingers closes and the elastic string is wound (Figure 2.30b). The cup with beans is then bound by the elastic string (Figure 2.30c). Finally, the binding hand picks up the bounded cup (Figure 2.30d). As shown in this demonstration, a binding hand can pick a paper cup with beans successfully. Note that the elastic string in contact with the cup applies normal and friction forces to the cup. Thanks to wider contact area between the elastic string and the cup, less stress is applied to the cup, reducing the deformation of the cup. Vertical component of applied normal and friction forces works as a lifting

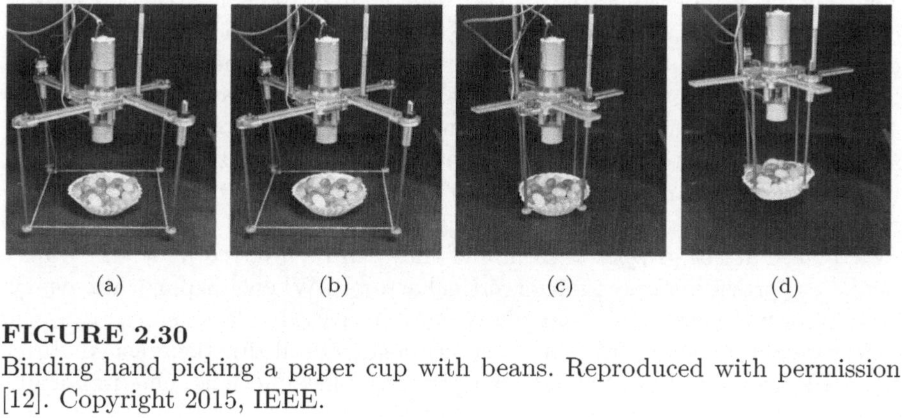

(a) (b) (c) (d)

FIGURE 2.30
Binding hand picking a paper cup with beans. Reproduced with permission [12]. Copyright 2015, IEEE.

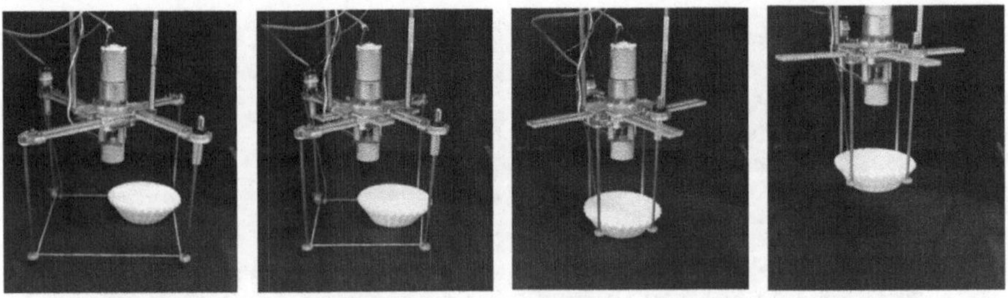

FIGURE 2.31
Centering a cup in the binding region. Reproduced with permission [12]. Copyright 2015, IEEE.

force. This lifting force is independent of the contact area, suggesting that we can obtain sufficient lifting force with less deformation of the cup.

Recall that manipulation process consists of approaching, grasping, moving, and releasing. Once a target object to be grasped is surrounded by an elastic string of a binding hand, the object is centered in the surrounded region as the hand closes its fingers. Figure 2.31 demonstrates this centering process. Initially, the cup filled with rubber was not located at the center of the binding region (Figure 2.31a). When binding started (Figure 2.31b), the elastic string was in contact with the cup. The contacting regions did not surround the cup, resulting in contacting force guiding the cup toward the center of the region. Finally, the contacting regions surrounded the cup (Figure 2.31c) and the binding hand picked up the cup (Figure 2.31d).

2.3.2 Scooping-binding hand

2.3.2.1 Concept

The binding gripper reported in [12] uses a single flexible string to form a closed loop for grasping and lifting object. However, binding grippers have difficulties in grasping thin, heavy, and slippery objects. Therefore, a scooping-binding gripper was proposed in [30] to handle such objects. The concept is shown in Figure 2.32a. A parallel configuration was adopted and multiple strings were used to provide larger grasping force (horizontal arrows) and friction force (shear direction arrows). Multiple strings can generate a 3D wrapping to better stabilize the grasping. Two thin plates (vertical thick lines) were placed at the gripper bottom and they can be inserted under the object bottom to provide support force (vertical arrows). When grasping thin, heavy, and slippery objects, the thin plates play a central role in supporting the object weight, and the strings are responsible for stabilizing the grasped object to avoid dropping during high-speed pick-and-place motions. One important

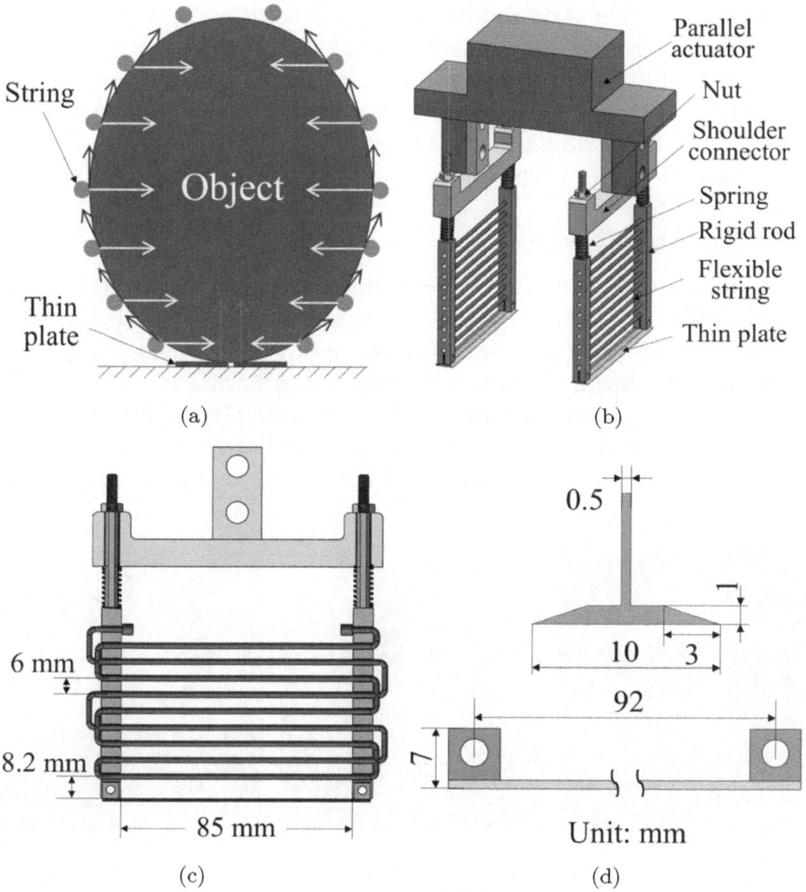

FIGURE 2.32
Concept and CAD designs of scooping-binding hand: (a) hand concept, (b) hand CAD design, (c) finger CAD design, and (d) scooping plate design. Reproduced with permission under the CC-BY 4.0 license [30].

feature of the scooping-binding gripper is that it requires active contact with a table or conveyor where the target object is placed on.

2.3.2.2 Gripper design

Design of the scooping-binding gripper is shown in Figure 2.32b. The gripper was constructed using a pneumatic parallel actuator and two identical scooping-binding mechanisms. This mechanism consists of a shoulder connector, two rigid rods, two springs, two nuts, a thin plate, and multiple strings between the two rigid rods. Given that the mechanism needs to make contact with the environment, two springs are employed to maintain contact and absorb the impact. The rigid rod has multiple holes to allow the string to pass through. The nut is used to connect the rod to the shoulder connector and adjust the preload of the spring. A slot is designed at the bottom end of the rod to assemble the thin plate. One single string was wound through both rods to position the string as shown in Figure 2.32c. The design of the thin plate is shown in Figure 2.32d.

2.3.2.3 Hand prototype

The shoulder connector, rigid rods, and thin plate were 3D printed with PLA (Polylactic acid) material. They were assembled together with the springs, nuts, and string to construct one scooping-binding mechanism (Figure 2.33a). The rubber string has a diameter of 3 mm and a hardness of Shore A55. A commercially available pneumatic parallel actuator (HLC-12AS-L1, Kondoh Mfg.Co.,Ltd., Nagoya) was used to realize the parallel open and close motions. The scooping-binding mechanisms were connected to the parallel actuator

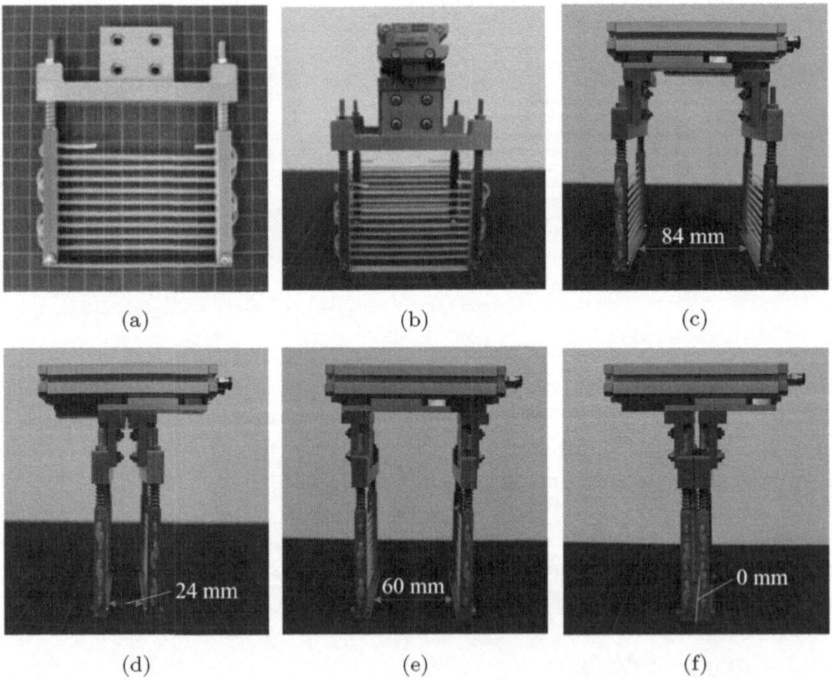

FIGURE 2.33
Manufactured prototype of scooping-binding hand: (a) scooping-binding mechanism, (b) hand prototype, (c) open state in wide configuration, (d) close state in wide configuration, (e) open state in narrow configuration, and (f) close state in narrow configuration. Reproduced with permission under the CC-BY 4.0 license [30].

using a base flange to complete the assembly of the scooping-binding gripper (Figure 2.33b). Depending on the fixture of the scooping-binding mechanisms, the gripper can have different widths upon opening and closing. For the wide configuration in which the scooping-binding mechanisms were connected to the external sides of the base flange, the gripper presented an opening width of 84 mm (Figure 2.33c) and a closing width of 24 mm (Figure 2.33d). By contrast, the gripper had a 60-mm (Figure 2.33e) opening and 0-mm (Figure 2.33f) closing widths for the narrow configuration in which the scooping-binding mechanisms were connected to the internal sides of the base flange.

2.3.2.4 Grasping experiments

Handling tests were conducted on various food items using a 4-DOF SCARA robot (HSR065A1-N, DENSO WAVE, Aichi). A total of 20 food items (Figure 2.34) were used in the experiments. The food items included fried foods (**B**, **C**,

FIGURE 2.34
Food items used in grasping experiments: (a) daifuku (small round mochi stuffed with sweet red bean paste), (b) fried shrimp, (c) ganmodoki (deep-fried tofu dumpling), (d) sweet potato tempura, (e) croquette, (f) tomato, (g) boiled flat fish cake, (h) sweet bean jelly, (i) rolled omelet, (j) green pepper, (k) kamaboko (steamed fish cake), (l) cheese, (m) octopus leg, (n) raw shrimp, (o) raw carangidae fish, (p) quail egg, (q) tofu, (r) meat, (s) raw oyster, and (t) Pollock roe. "SR" indicates the success rate of the handling tests. Reproduced with permission under the CC-BY 4.0 license [30].

D, **E**), very soft foods (**A**, **I**, **Q**, **S**, **T**), relatively heavy foods (**A**, **F**, **Q**), fragile foods (**S**, **T**), foods with thin profiles (**D**, **E**, **G**, **H**, **N**, **O**, **S**, **T**), and food with slippery surfaces (**G**, **M**, **N**, **O**, **S**, **T**). Some of them, such as the raw oyster (**S**) and the Pollock roe (**T**), combine several properties that are difficult for robotic grippers to manage.

The handling test is a conventional pick-and-place operation. The experimental system is shown in Figure 2.35a. The scooping-binding gripper was attached to the SCARA robot using a 3D printed flange interface. A conveyor was used to place food items. The gripper picked the food item and placed it into a stainless tray. The maximum operation speed of the robot was used to test the stability of grasping. Five trials were conducted for each food item, and the posture of the food for each trial was manually adjusted to account for an arbitrary grasping posture.

Some experimental snapshots while grasping various food items are shown in Figures 2.35b–i. The gripper succeeded in handling all food items with a success rate of 100% except for the raw shrimp (80%), the carangidae fish (60%), and the raw oyster (80%). The results demonstrate that the scooping-binding gripper could handle all food items despite their complex geometries, thin profiles, slippery surface conditions, and variations in shape, size, and

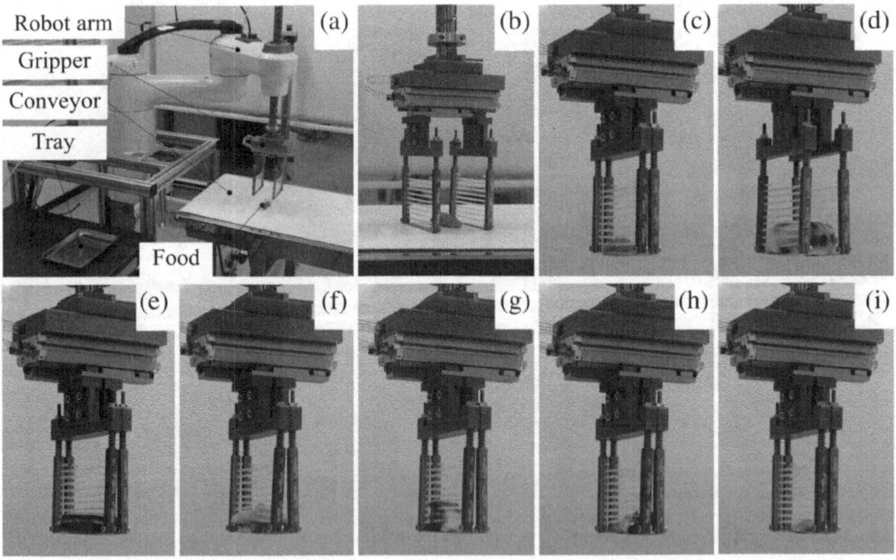

FIGURE 2.35
Experimental snapshots of food-handling tests: (a) Experimental setup of the handling test. (b) Opening state before grasping. Experimental snapshots while grasping and lifting the (c) pollock roe, (d) daifuku, (e) sweet bean jelly, (f) sweet potato tempura, (g) octopus leg, (h) carangidae fish, and (i) raw oyster, respectively. Reproduced with permission under the CC-BY 4.0 license [30].

softness. For handling daifuku, tomato, tofu, and meat, a wide configuration of the gripper was used because of the size of the products.

The gripper failed one trial for handling raw shrimp and raw oyster, respectively, and two trials for handling the carangidae fish. This was due to the slippery property and minimum height profiles of these food items. Their maximum heights are higher than the gap between the lowest rubber string and the thin plate, but their minimum heights are smaller than that gap. At a certain grasping orientation, the food item may slip out through the gap due to slippery surface. This problem can be solved by reducing the gap between the lowest string and the thin plate and increasing the friction coefficient of the string and the thin plate.

The takt time of the handling tests was approximately 4 s, and the results validated the stability of the scooping-binding gripper working at high-speed motion. During experiments, significant damage to the food items was not found because of the flexibility of rubber string. The takt time depends on the travel distance of the pick-and-place motion and it can be further reduced by using a faster robot manipulator, such as a parallel robot. Multi-gripper system being able to handling multiple food products at once is another solution to shorten the takt time.

2.4 Circular Membrane Hand

2.4.1 Concept

The motivation behind this work arises from the need for high-speed handling of fragile foods and fruit harvesting. High-speed handling involves grasping, lifting, and translation motions, as illustrated in Figure 2.36a. In contrast, harvesting requires grasping and pulling motions with an arbitrary handling posture, in addition to a twisting motion for successful execution, as shown in Figure 2.36b.

Traditional multi-finger soft hands typically grasp an object at the fingertips (Figure 2.36c), which results in a small contact area and, therefore, requires a large grasping force to ensure sufficient friction for stable handling. To increase the contact area, wider fingers could be employed, as depicted in Figure 2.36d. While this does increase the contact area and helps distribute the grasping force, soft fingers still struggle to guarantee stable handling during high-speed motions due to inertia forces.

To address this, a hybrid circular membrane hand was proposed in [31,32], as shown in Figure 2.36e. This design consists of a rigid shell with multiple soft chambers inside. When the soft chambers are inflated, they generate a large contact area, reducing the risk of damaging the product. The rigid shell provides the necessary rigidity to withstand inertia forces during high-speed motion and enables arbitrary handling postures and twisting motions.

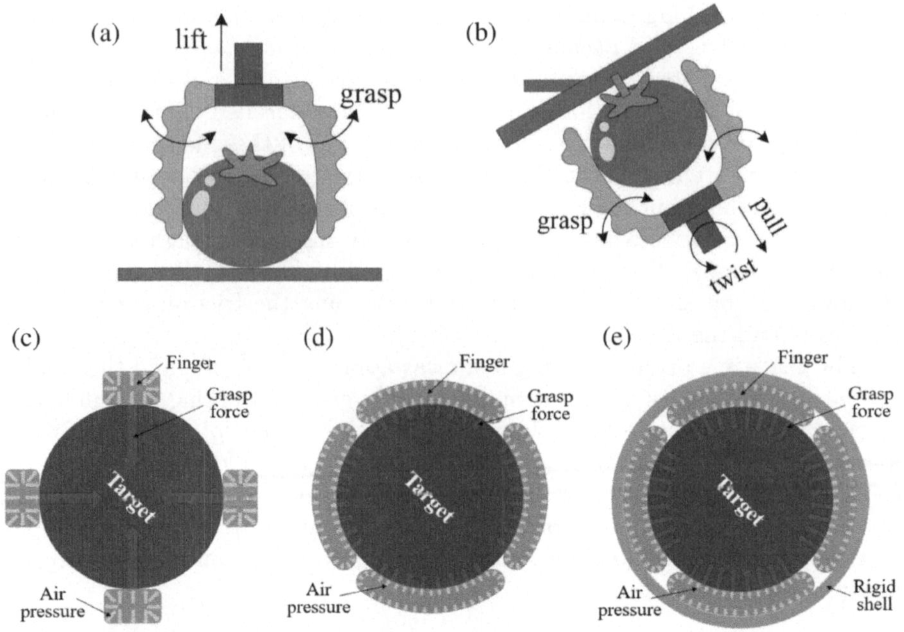

FIGURE 2.36
Concept of circular membrane hand: (a) soft hand lifting a food item with grasping and lifting motions, (b) soft hand harvesting an agricultural product with grasping, twisting, and pulling motions, (c) hand with four narrow soft fingers, (d) hand with four wide soft fingers, and (e) proposed hand with four wide soft fingers and a rigid external shell.

2.4.2 Hand design and fabrication

The hand was designed in cylindrical shape, where the internal diameter of the soft chambers is 81 mm and the height is 40 mm to accommodate potential variations in geometrical dimensions. The design is divided into four identical modules, each comprising a rigid shell and a soft chamber, occupying one-quarter of a circle (Figure 2.37a). The air chamber has a total thickness of 9.5 mm, with a wall thickness of 1.5 mm. The overall thickness of each module, including the rigid shell, is 14.5 mm. The circular membrane hand is assembled by combining these four identical modules and attaching a rigid frame, as shown in Figure 2.37b. This modular design facilitates easy fabrication.

The soft chambers were cast using two sets of molds (Figure 2.38a), which were 3D printed using ABS material. The mixed liquid rubber was first poured into the chamber mold after degassing to cast an open chamber. At the same time, a cover was cast using a separate mold. After curing, the open chamber and cover were demolded and bonded together using the same liquid rubber. This process was relatively straightforward. The resulting soft chamber is

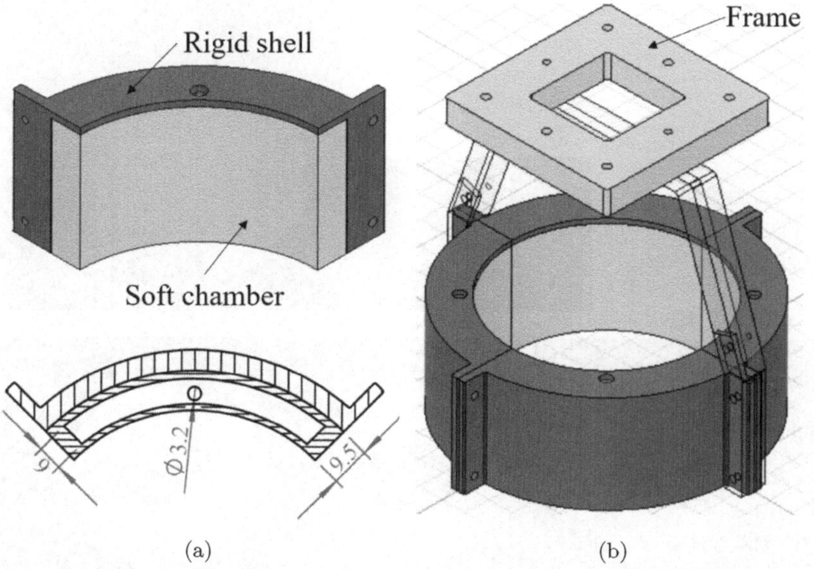

FIGURE 2.37
Design of circular membrane hand: (a) components of a hand module and (b) hand assembly with four modules. Reproduced with permission [32]. Copyright 2021, Mary Ann Liebert, Inc.

shown in Figure 2.38b. The rigid shell was also 3D printed using ABS material. The soft chamber and shell were then glued together using an instant adhesive (Cemedine Co., Ltd., Tokyo) to complete the fabrication of a single module (Figure 2.38c). The circular membrane hand was assembled by combining four such modules. For comparison, three hands (G1, G2, and G3 in Figure 2.38d) were fabricated using three different soft materials – Dragon Skin 10, 20, and 30 (Smooth-On, Inc., PA), which have Shore hardness values of 10A, 20A, and 30A, respectively.

2.4.3 Force characterization

The experimental setup is shown in Figure 2.39. A cylinder, 3D printed with ABS material and having a diameter of 69 mm, was mounted on a six-axis force sensor (PFS055YA251U6, Leptrino Inc., Japan). The circular membrane hand was attached to a robotic manipulator (VP-6242, DENSO WAVE Inc., Japan). The robotic manipulator was programmed to perform grasp-and-pick and grasp-and-twist motions, while the lifting force and twisting torque were measured by the force sensor. The motions were executed until slippage occurred, with the maximum force and torque recorded as the lifting force and twisting torque, respectively. The three hands shown in Figure 2.38d were tested at two different pressures, 7 kPa and 10 kPa, for comparison. The lifting

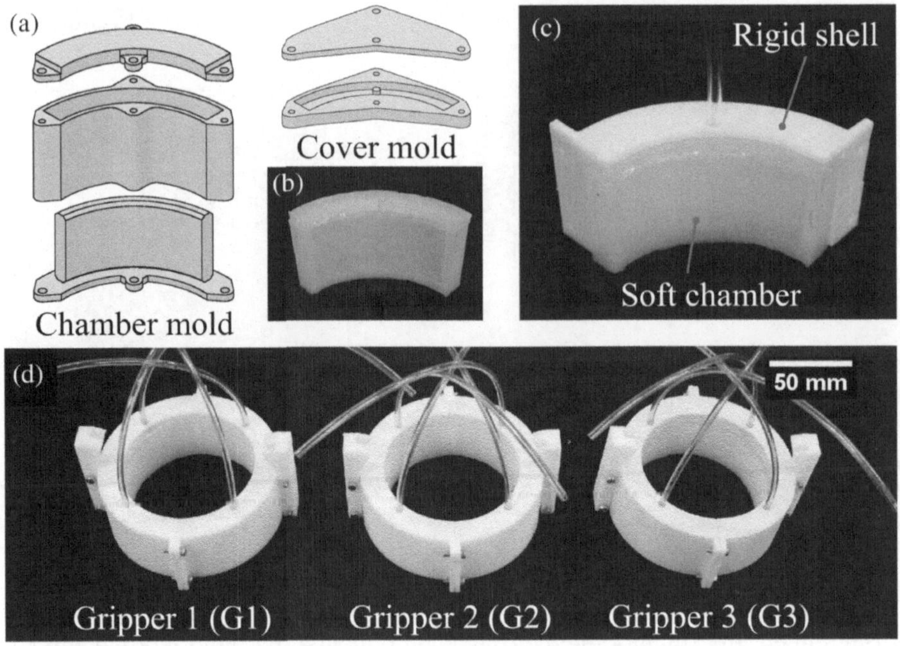

FIGURE 2.38
Fabrication of circular membrane hand: (a) molds for fabricating the soft chamber, (b) fabricated soft chamber, (c) single module with rigid shell and soft chamber, and (d) the three fabricated hands. The soft chambers are fabricated with silicone material of Dragon Skin 10 (DS10), Dragon skin 20 (DS20), and Dragon Skin 30 (DS30), respectively. Reproduced with permission [32]. Copyright 2021, Mary Ann Liebert, Inc.

force tests were repeated ten times for each pressure value, while the twisting torque tests were repeated five times in both counterclockwise and clockwise directions.

The measured lifting force and twisting torque are summarized in Table 2.1, where "End" and "Max" indicate the force or torque values at the stable state and maximum moment, respectively. The hand G1 generates a larger force and torque than the other two hands at the same pressure. This behavior was more pronounced at a lower input pressure. This is because that the material of G1 is softer and more stretchable. Therefore, it can generate larger contact area than the other two hands with harder materials.

2.4.4 Grasping tests on various targets

A total of ten food and drink-related targets (Figure 2.40) and a SCARA robot (HSR065A1, DENSO WAVE Inc., Japan) were used in the tests. A pick-and-place motion was programmed, and different time delays were investigated

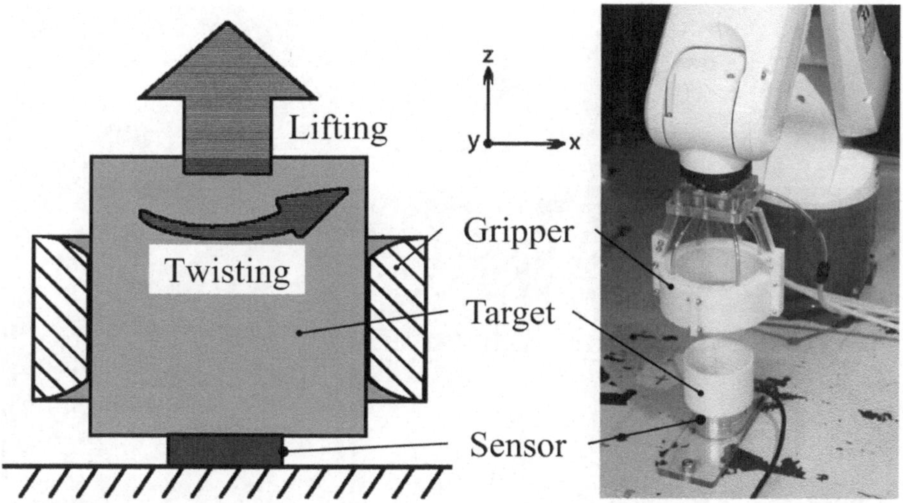

FIGURE 2.39
Experimental setup for force characterization. Reproduced with permission [32]. Copyright 2021, Mary Ann Liebert, Inc.

for picking and placing various targets. During these time delays, the robot sent on-off signals to control a solenoid valve (VQ110U-5M-M5, SMC, Tokyo) for inflating and deflating the hand. These time delays were referred to as the "pick time" and "place time", and the objective of these tests was to determine the optimal time delays for successfully completing the handling tasks. The successful trials were repeated five times for confirmation.

The air pressure was predetermined based on the size of the target. Preliminary tests showed that the smallest size the hands could successfully grasp was approximately 45 mm in diameter. Consequently, for handling a raw egg with a diameter of 42.5 mm, a smaller hand (G-small) was fabricated with dimensions of 46 mm (internal diameter) × 72 mm (external diameter) × 30 mm (height), using Dragon Skin 10. To further optimize takt time, a parallel

TABLE 2.1
Lifting force and twisting torque

	Lifting force [N]				Twisting torque [Nm]			
	P = 7 kPa		P = 10 kPa		P = 7 kPa		P = 10 kPa	
Material	End	Max	End	Max	End	Max	End	Max
DS10	37.51	40.72	48.22	50.97	0.63	0.64	0.71	0.73
DS20	9.27	15.16	25.41	35.07	0.29	0.38	0.53	0.71
DS30	3.96	7.05	9.74	18.00	0.12	0.18	0.25	0.39

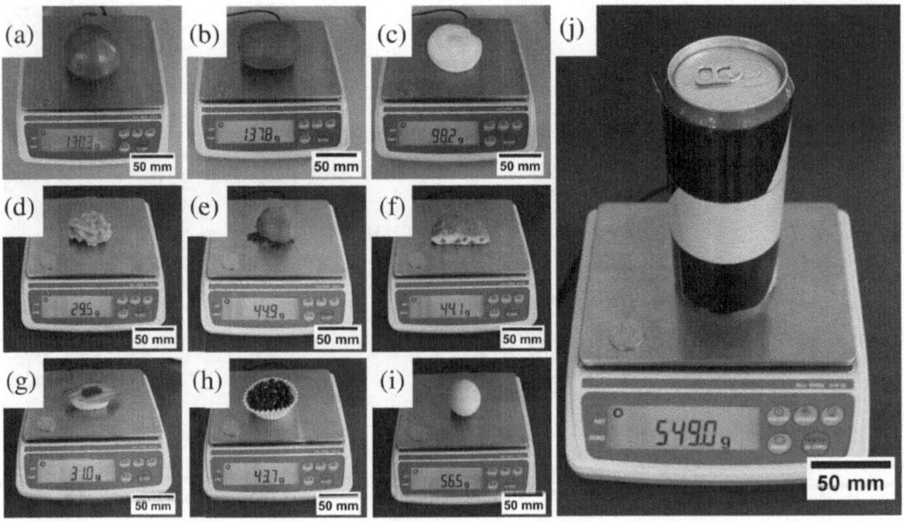

FIGURE 2.40
Objects used in handling tests: (a) a tomato (130.3 g), (b) a kiwifruit (137.8 g), (c) a Daifuku (98.2 g), (d) a fried chicken (29.5 g), (e) a strawberry (44.9 g), (f) half cut of a potato croquette (44.1 g), (g) dummy sample of a sushi (31.0 g), (h) paper cup filled with red beans (43.7 g), (i) a raw egg (56.5 g), and (j) a canned drink (549.0 g). Reproduced with permission [32]. Copyright 2021, Mary Ann Liebert, Inc.

robot (IRB 360 FlexPicker, ABB, Switzerland) was employed to handle the egg, operating at a maximum speed of 10,000 mm/s.

The experimental results for handling different targets are presented in Table 2.2. Results for the potato croquette (2.40F) are not listed, as all hands failed to grasp it. The hand G1 successfully handled all other targets, while hands G2 and G3 failed to handle the strawberry and dummy sushi. This failure can be attributed to the higher stretchability of Dragon Skin 10 compared to the other two materials. The hand G1 was able to inflate sufficiently at relatively low pressures to enclose the targets. However, it required a longer inflation and deflation time, resulting in a slightly longer takt time. As a result, a takt time of approximately 2–5 s was achieved for handling most targets using the SCARA robot. With a parallel robot, the takt time was reduced to under 1 s for the raw egg test.

The circular membrane hand demonstrated robustness in handling circular or nearly circular targets, such as tomatoes, kiwifruits in a vertical orientation, Daifuku, canned drinks, and eggs. It also successfully handled deformable targets like Daifuku and a paper cup filled with red beans. Additionally, the hand

TABLE 2.2
Experimental results of handling tests

Object	hand	Pressure [kPa]	Pick time [s]	Place time [s]	Takt time [s]
Tomato	G1	10	0.7	0.7	3.75
	G2	20	0.5	0.3	3.15
	G3	30	0.4	0.5	3.25
Kiwifruit (vertical)	G1	20	0.8	0.8	3.95
	G2	30	0.9	0.2	3.45
	G3	30	1.0	0.2	3.55
Kiwifruit (horizontal)	G1	10	0.6	0.6	3.55
	G2	20	0.2	0.6	3.15
	G3	30	0.2	0.7	3.25
Daifuku	G1	20	1.0	1.0	4.35
	G2	30	1.1	0.1	3.55
	G3	30	1.4	0.1	3.85
Fried chicken	G1	20	1.5	0.8	4.65
	G2	30	0.8	0.2	3.35
	G3	30	1.0	0.2	3.55
Strawberry	G1	10	1.8	1.0	5.15
	G2	30	×	×	×
	G3	30	×	×	×
Dummy sushi	G1	20	1.6	2.3	6.25
	G2	30	×	×	×
	G3	30	×	×	×
Paper cup with beans	G1	10	1.3	0.6	4.25
	G2	30	0.7	0.3	3.35
	G3	30	0.8	0.2	3.35
Canned drink	G1	10	0.6	0.6	3.55
	G2	30	0.3	0.5	3.15
	G3	30	0.4	0.4	3.15
Raw egg	G-small	10	0.1	0.1	2.55
Parallel	G-small	10	0.1	0.1	0.96

Note: For kiwifruit, "vertical" indicates that the long axis of the kiwifruit is vertical to the ground, and "horizontal" indicates that the axis is parallel to the ground. "Parallel" indicates the trial using a parallel robot. Symbol "×" indicates a failed test.

was capable of handling oval-shaped targets, such as kiwifruits in a horizontal orientation, strawberries, and dummy sushi. However, the hands struggled with low-profile targets like the potato croquette. The inflated chambers exerted a downward pressing force on low-profile objects, which was insufficient to overcome the combined downward force and gravity, preventing successful lifting. This limitation in handling low-profile objects is a key drawback of the current circular membrane hand design.

2.4.5 Object manipulation tests

To demonstrate the circular membrane hand's capability of manipulating objects in arbitrary orientations and performing diverse tasks, two object manipulation tests were conducted. The first test involved swaying a tomato and a canned drink in an inclined orientation. A pick-and-sway motion was programmed into the robot manipulator (UR5e, Universal Robots, Denmark) at 100% operation speed. After lifting the target, the robot tilted it into an inclined position and then performed a sway motion from left to right before returning the target to its original position. Experimental snapshots are shown in Figure 2.41. The results indicated that the circular membrane hand had no difficulty handling the targets in an inclined orientation during high-speed sway motions.

The second test involved pouring wine into a glass (Figure 2.42). A half bottle of wine was placed on a wooden wine stand in an arbitrary posture. First, the circular membrane hand was driven to approach the bottle and grasp its bottom. Then, the robot tilted the bottle upward, with the author assisting by holding the bottle cap. Next, the circular membrane hand performed a twisting motion to uncork the bottle. Finally, the hand holding the bottle was moved toward the wine glass, and a pouring motion was executed. This test demonstrated that the circular membrane hand is capable of accommodating arbitrary grasping postures and performing twisting motions.

(a) (b)

FIGURE 2.41
Manipulating (a) a tomato and (b) a juice can in arbitrary orientation. Reproduced with permission [32]. Copyright 2021, Mary Ann Liebert, Inc.

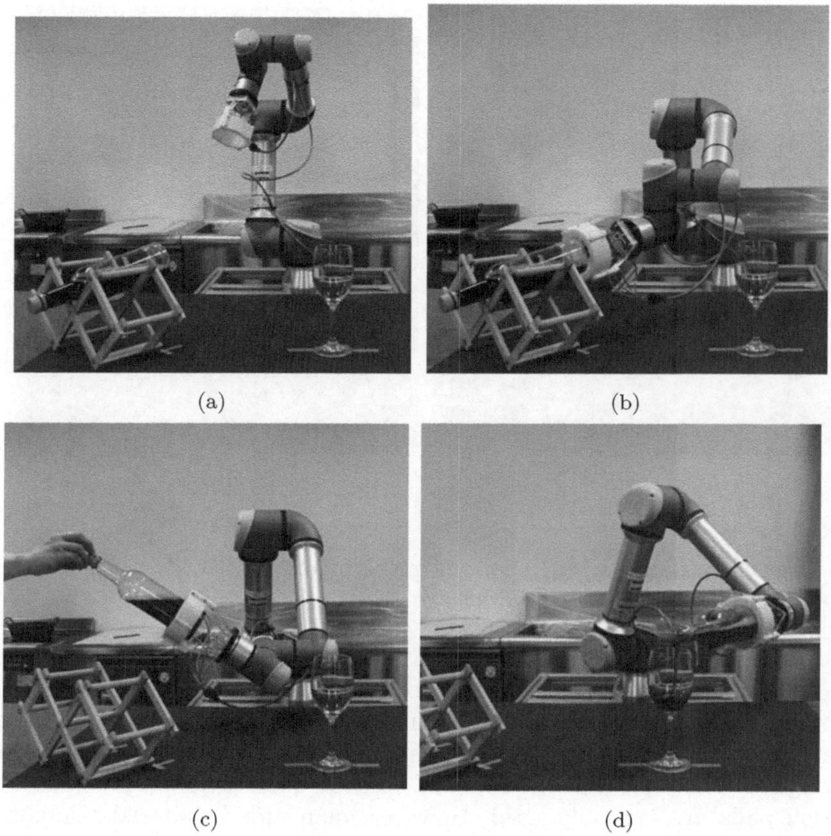

FIGURE 2.42
Pouring wine test using circular membrane hand: (a) initial position, (b) approaching wine bottle, (c) uncapping wine bottle, and (d) pouring wine into glass. Reproduced with permission [32]. Copyright 2021, Mary Ann Liebert, Inc.

2.5 Needle Hand

Despite the advancements in robotic hands, handling shredded and chopped food materials (see Figure 2.43a) and food products packaged inside tray slots (see Figure 2.43b) remains a challenge for current robotic hands. In the scenario shown in Figure 2.43a, the robotic hand needs to have a certain level of rigidity and a thin structure to successfully insert its fingers into the food material for grasping. It also must be able to hold the grasped food without dropping it. The situation depicted in Figure 2.43b typically arises during

(a) (b)

FIGURE 2.43
Food material examples: (a) chopped green onion and (b) deep fried shrimp. Reproduced with permission under the CC BY-ND 4.0 license [34].

production inspection operations. If a defective product is identified on the production line, it must be removed from its packaged slot. Currently, such inspection tasks are performed by human workers. Since the food product is located inside a confined slot, the lateral space around it is extremely limited, leaving only the top surface of the food accessible for grasping. In these cases, suction pads are typically used. However, many food materials cannot be grasped by suction pads due to their surface properties. To address this issue, a needle hand was proposed in [33, 34].

2.5.1 Concept

The concept of the needle hand is illustrated in Figure 2.44. The hand is constructed with multiple needles and a soft membrane. The membrane, in combination with the hand base, forms an air chamber. When pressure or vacuum is applied to this chamber, the membrane either inflates or deflates. In the grasping mode (see Figure 2.44a), the membrane is initially inflated to open the needles. The hand can then be inserted into shredded or chopped food materials. Once insertion is complete, a vacuum is applied to deflate the membrane and close the hand around the material. The thin structure of the needles allows the hand to easily penetrate various types of shredded or chopped food.

In the piercing mode (see Figure 2.44b), the needle hand is driven to pierce the food material, with the needles oriented vertically downward to facilitate the piercing process. Once piercing is completed, air or vacuum pressure is applied to inflate or deflate the membrane, generating a holding force that

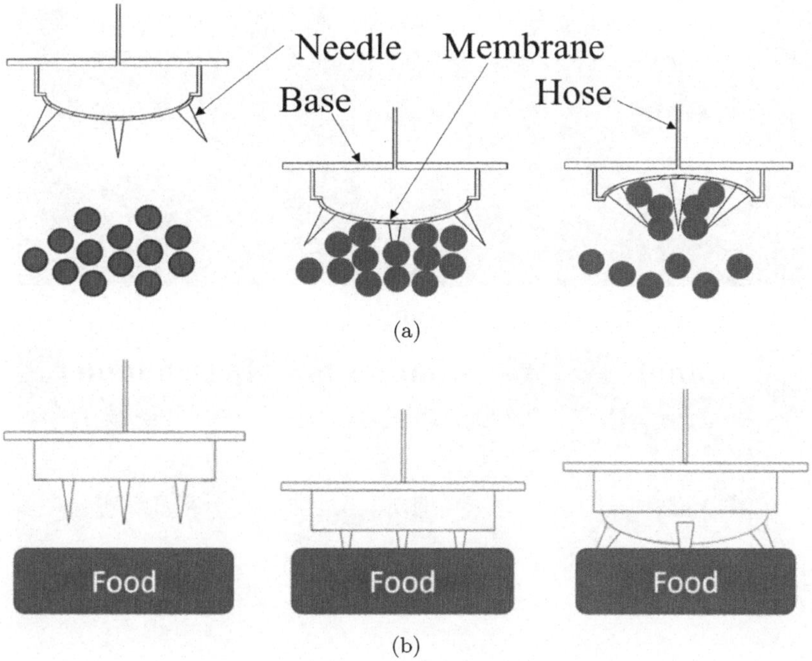

FIGURE 2.44
Concept of needle hand in (a) grasping and (b) piercing modes. Reproduced with permission under the CC BY-ND 4.0 license [34].

secures the food material. The needle hand features a simple structure and can be operated using a straightforward control strategy.

2.5.2 Needle hand prototype

The grippers were fabricated using a multi-material 3D printer (Objet350 Connex3, Stratasys, MN). The hand base and needles were 3D printed using VeroWhite (a rigid material), while the soft membrane was printed with Agilus30 (a rubber-like material with a Shore hardness of A30-35 and elongation at break of 220–270%). An air hose was inserted into the hand base to apply air pressure. Examples of the fabricated hands in different states for grasping and piercing operations are shown in Figure 2.45.

For the grasping needle hand, needles were initially closed, with the tips touching each other (left side of Figure 2.45a). When air pressure was applied, the gripper opened (middle of Figure 2.45a), allowing the hand to grasp. Applying vacuum pressure caused the gripper to close, with the needles crossing each other (right side of Figure 2.45a).

For the piercing hand, needles were oriented downward in the initial state (left side of Figure 2.45b). The gripper can be inflated and deflated to generate

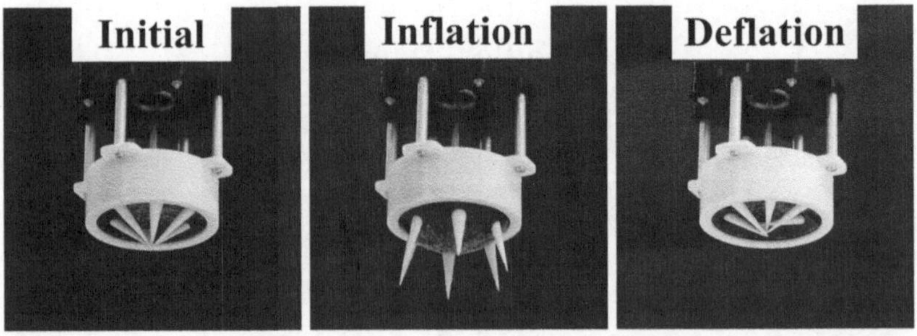

(a)

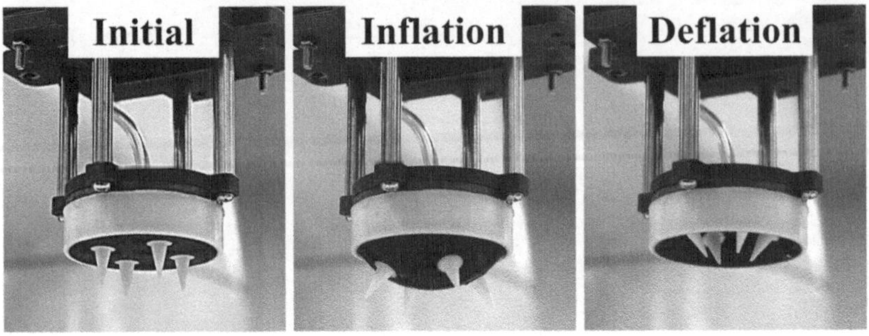

(b)

FIGURE 2.45
Manufactured prototypes of (a) grasping and (b) piercing needle hands. Reproduced with permission under the CC BY-ND 4.0 license [34].

the opening (middle of Figure 2.45b) and closing (right side of Figure 2.45b) states, respectively.

2.5.3 Grasping tests

The food materials used in the grasping tests included shredded cabbage, salt kelp, hijiki, and green onions chopped into two different sizes, as shown in Figure 2.46. These materials were prepared in containers and manually flattened for tests. The containers had varying dimensions: 350 mm × 260 mm × 45 mm (see Figures 2.46a and 2.46b), ϕ320 mm × 120 mm (see Figure 2.46c), and 340 mm × 260 mm × 80 mm (see Figure 2.46d). The food materials were purchased online and immediately stored in a commercial refrigerator upon arrival. All experimental tests were conducted on the same day the materials were received.

The experimental setup for grasping tests is shown in Figure 2.47a. The needle hand was mounted on a SCARA robot arm (HSR065A1, DENSO WAVE Inc., Aichi), which was programmed to perform pick-and-place operations. An electric scale was used to measure the weight of the grasped

FIGURE 2.46
Food materials used in the grasping tests: (a) chopped cabbage, (b) salt kelp, (c) hijiki, and (d) chopped green onion. Reproduced with permission under the CC BY-ND 4.0 license [34].

food. Air pressures of 30 kPa and −30 kPa were applied to open and close the hand, respectively. Six grasps were performed in each experimental trial, with the circles shown on the left of Figure 2.47b indicating the positions where the six items were placed. Five experimental trials were conducted for each food material, resulting in a total of 30 grasps per material. The weights and appearances (captured via camera images) of all the grasps were recorded for further analysis.

To evaluate the performance of the grasping tests, two measures were investigated: weight and appearance. Weight data was obtained using an electric scale during the experimental tests, and it could also be easily measured by incorporating a force sensor into the production line of a food factory. However, appearance is another critical factor in the packaging or topping of shredded and chopped food materials, as it relates to how much area or volume is occupied for visual satisfaction.

For appearance, the occupied area of the grasped food material was calculated. In actual food production lines, cameras are commonly available, making it relatively easy to capture images of the grasped food. To measure the appearance, each image was converted to a binary format (see the right of Figure 2.47b). The total number of pixels corresponding to the black region was then calculated, which was considered the occupied area of the grasped food material.

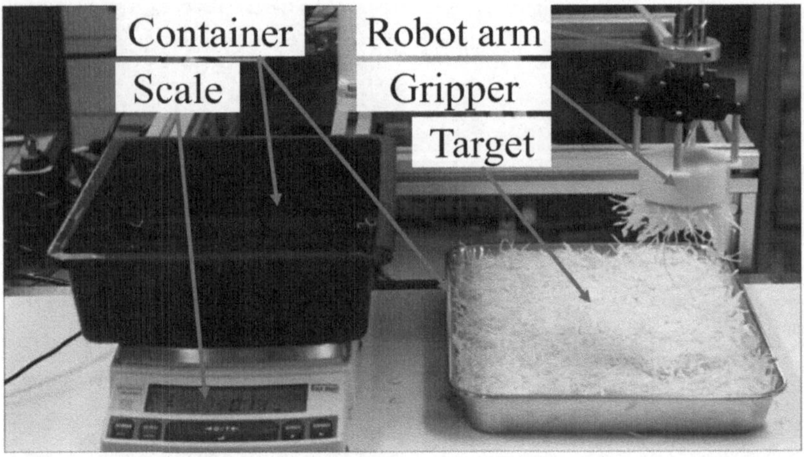

(a)

(b)

FIGURE 2.47
Food grasping experiments of needle hand: (a) experimental setup and (b) place result and binarized image. Reproduced with permission under the CC BY-ND 4.0 license [34].

The results of the averaged weights and the total number of pixels for different food materials are shown in Figure 2.48. Two needle hands, one with 6 needles and the other with 12 needles, were used in the tests for comparison. The error bars represent the standard deviations from the 30 grasps.

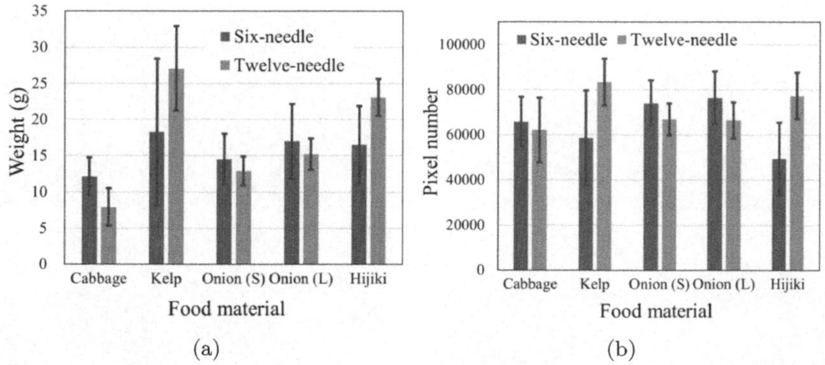

FIGURE 2.48
Evaluation of grasping experiments: (a) average weight and (b) pixel number. Reproduced with permission under the CC BY-ND 4.0 license [34].

Compared to the 6-needle hand, the 12-needle hand was able to grasp larger amounts of salt kelp and hijiki, but smaller amounts of the other three food materials. This variation can be attributed to the different physical properties of food materials. Salt kelp and hijiki tend to scatter more easily and are more prone to dropping during lifting, while cabbage and onions are more entangled, meaning that fewer needles create more space for accommodating larger quantities of food.

The 12-needle hand also exhibited less variability, with smaller standard deviations compared to the 6-needle hand. This suggests that the 12-needle hand offers more stable grasping performance. Furthermore, the appearance measure (Figure 2.48b) showed less variation among the different food materials compared to the weight measure (Figure 2.48a). This indicates that the grasped food materials may occupy a similar area, even though their weights differ.

2.5.4 Piercing tests

Piercing tests were conducted on 11 food materials (see Figure 2.49). To minimize variations in shape and properties of each food material, bagged frozen food products were purchased. Prior to the experiments, the food materials were defrosted and cooked in a microwave oven according to the product recipes.

The experimental setup for the piercing tests is shown in Figure 2.50. As in the grasping tests, the gripper was mounted on the same SCARA robot. The robot arm translated the robotic hand downward to pierce the food product. After piercing, air pressure was applied to the hand to generate a holding force inside the food product. For each experimental trial, two samples of each food product were tested successively using a pick-and-place motion. If the hand

FIGURE 2.49
Food items used in the piercing tests: (a) grilled salmon, (b) hamburger, (c) croquette, (d) nugget, (e) tonkatsu, (f) meat cutlet, (g) tuna tatsuta, (h) fried fish, (i) sesame dumpling, (j) omelette, and (k) sweet potato pie. Reproduced with permission under the CC BY-ND 4.0 license [34].

failed to release the first food product at the placement position, the test was manually stopped, and the food product was removed. Then, the test was started for the second food product.

The results of the piercing tests are presented in Table 2.3. The needle hand failed to pick up the omelette due to its softness. The hand successfully picked up the tonkatsu and sesame dumpling, but failed to release them at the placement position. Unlike other fried meat products, such as hamburgers, nuggets, and meat cutlets – which are made from meat surimi (paste) – tonkatsu is made by deeply frying meat that contains harder fibers. The needles may have become stuck between these hard meat fibers, potentially causing the failure to release the food. Additionally, the sesame dumplings have a sticky mochi covering, which likely adhered to the needles, preventing proper release.

To investigate the influence of applied air pressure during the holding phase, piercing tests were conducted on two food materials, tonkatsu and

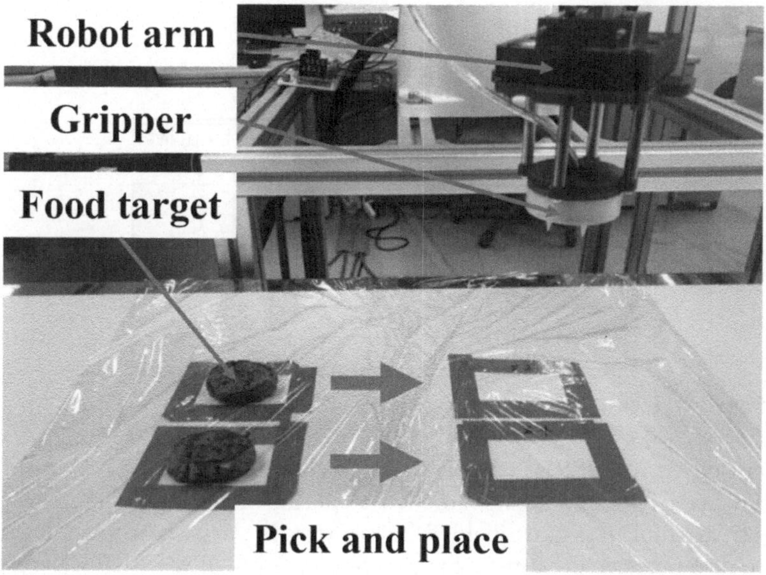

FIGURE 2.50
Experimental setup for the piercing tests. Reproduced with permission under the CC BY-ND 4.0 license [34].

TABLE 2.3
Results of the piercing tests

Food	Size [mm]	Mass [g]	1st	2nd
A	40 × 55 × 20	22.54	○	○
B	20 × 55 × 20	20.08	○	○
C	50 × 50 × 30	30.18	○	○
D	45 × 45 × 25	22.50	○	○
E	40 × 45 × 20	21.49	△	△
F	40 × 45 × 20	23.22	○	○
G	40 × 50 × 15	15.67	○	○
H	40 × 55 × 20	23.05	○	○
I	40 × 40 × 40	20.58	△	△
J	35 × 65 × 20	25.11	×	×
K	30 × 60 × 40	15.08	○	○

Note: "○", "△", and "×" indicate success in both picking and releasing, success in picking but failure in releasing, and failure in picking, respectively. The labels (A, B, C, ...) for food items relate to Figure 2.49.
Reproduced with permission under the CC BY-ND 4.0 license [34].

TABLE 2.4
Results of piercing tests at different pressures

	Tonkatsu		Meat cutlet	
Pressure [kPa]	1st	2nd	1st	2nd
10	△	△	○	○
20	△	△	○	○
30	△	△	○	○
40	△	△	○	○
−20	○	○	○	○
−30	○	○	○	○

Note: Reproduced with permission under the CC BY-ND 4.0 license [34].

meat cutlet, at six different pressures (see Table 2.4). The results indicated that varying pressures did not affect the performance with the meat cutlet; however, vacuum pressures yielded better results with the tonkatsu material compared to positive pressures. These differences are illustrated in Figure 2.51.

When positive pressures were applied during holding, the membrane inflated, generating an outward force on the grasped food material (see Figure 2.51a). This outward force caused the needles to penetrate deeper into the food, making it more difficult to release the food. In contrast, when vacuum pressure was applied, the membrane deflated, generating an inward force on the food (see Figure 2.51b). This inward force helped to pull the needles out of the food material, facilitating the release of needles upon placing the food.

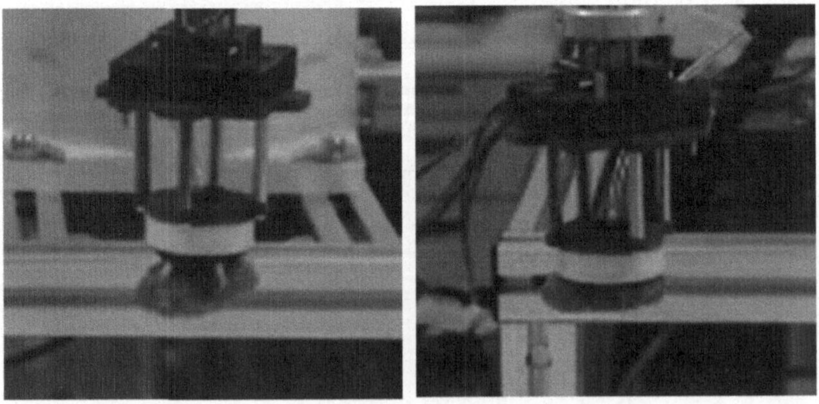

(a) (b)

FIGURE 2.51
Experimental snapshots of piercing tests: (a) applying positive pressure after piercing and (b) applying vacuum pressure after piercing. Reproduced with permission under the CC BY-ND 4.0 license [34].

3

Robotic Hands for Agriculture

This chapter introduces robotic hands for manipulation in agricultural applications. Two robotic hands are presented in this chapter. One is the planar membrane hand for automatic packaging of multiple cucumbers simultaneously. The other is the parallel robotic hand actuated using soft bellows actuators for handling agricultural products, such as radish, potato, carrot, and burdock root.

3.1 Planar Membrane Hand

3.1.1 Concept

To firmly grasp an object and handle it at high speeds, the concept of a membrane hand was proposed. This hand can be arranged in either a circular or planar configuration, as shown in Figure 3.1. As the name suggests, the membrane hand is constructed using rigid shells and soft membrane chambers. The soft membrane is bonded to the rigid shell, and when the membrane is inflated, the hand can securely grasp an object. Due to the presence of the rigid shell, the membrane hand can generate larger forces than entirely soft hands, providing more stability during grasping and enabling high-speed operation.

The circular membrane hand (see the left of Figure 3.1) is particularly effective for handling objects with an approximately circular shape, as discussed in [32]. In contrast, the planar membrane hand (see the right of Figure 3.1) is better suited for handling elongated objects, such as cucumbers. This chapter presents and discusses the results related to parallel membrane hand.

3.1.2 Fabrication of planar membrane hand

Generally, the casting method is employed to fabricate soft robots using silicone rubber materials. The casting process typically involves several steps, including mixing two liquid components, degassing the mixture, pouring it into molds, performing a second degassing (though this may not always be necessary), curing the silicone, demolding, and occasionally gluing the silicone

DOI: 10.1201/9781003372431-3

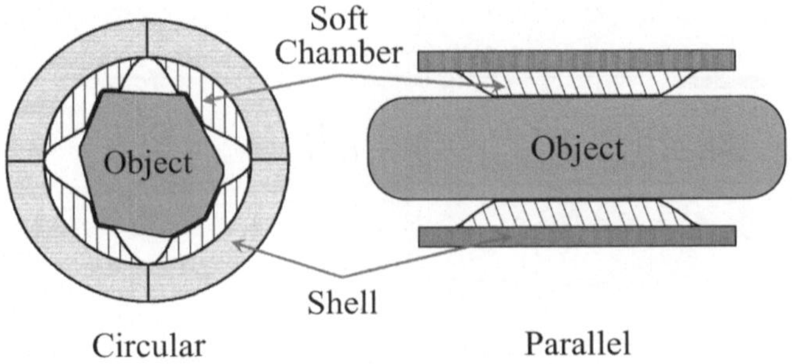

FIGURE 3.1
Concept of planar membrane hand. Reproduced with permission [31]. Copyright 2020, IEEE.

parts together [31,32]. Among these, the demolding and gluing operations are particularly critical, as the quality of the soft robot is highly dependent on their execution. Improper gluing can often lead to air leakage. To address this issue, the work in [35] proposed a demolding-free casting method to eliminate the need for both demolding and gluing operations when fabricating the planar membrane hand.

The internal structure of the planar membrane hand, which can be fabricated using the demolding-free casting method, is shown in Figure 3.2a. The cross-section of this structure consists of three parts: shell, silicone, and air path. By designing the shell as illustrated in Figure 3.2a, silicone rubber can be cast around the shell, preventing the silicone membrane from peeling off.

Three rigid components are necessary to fabricate the planar membrane hand: two shell parts and a hose insert, which is used to block the air path during the curing of silicone rubber. The fabrication process is outlined in Figure 3.2b. First, the two shell parts, 3D-printed using an Original Prusa i3 MK3S (Prusa Research a.s., Czech Republic), are glued together. Next, liquid silicone is poured into the bonded shell and cured at room temperature. Finally, by inserting an air tube into the air path, the fabrication is completed, resulting in a hand with a total thickness of just 3 mm. Since the shell also serves as the casting mold, this process eliminates the need for a demolding operation, thus making it a "demolding-free" method.

3.1.3 Inflation characterization

The displacement of the inflated membrane was measured to evaluate the performance of the planar membrane hand. In this test, planar hands fabricated with two different silicone materials – Dragon Skin 20 and Dragon Skin 30 (Smooth-On, PA) – were compared. For a parallel configuration with a 40 mm

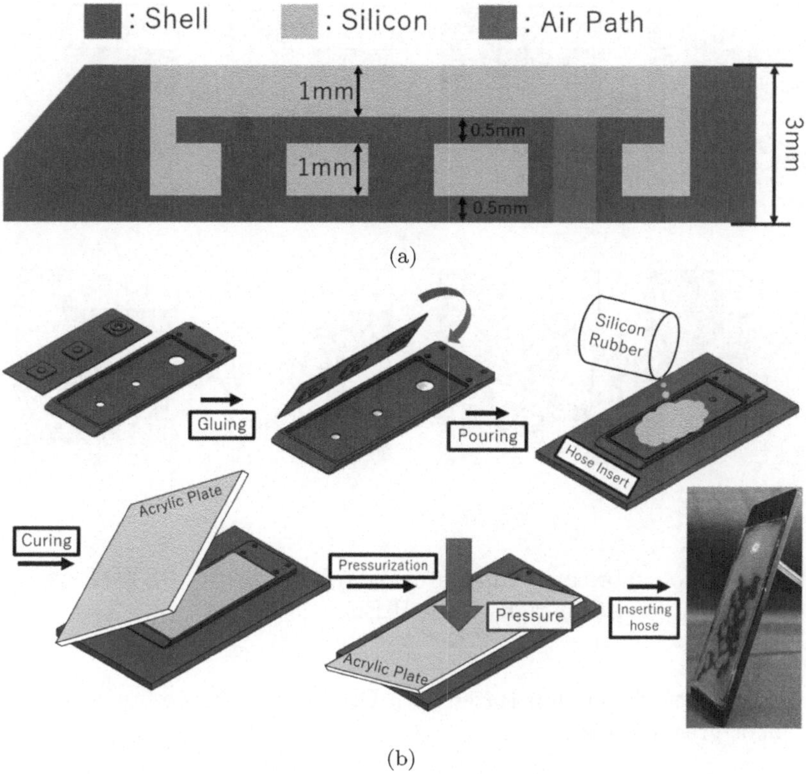

FIGURE 3.2
Finger design and fabrication. (a) Internal structure of planar membrane hand. (b) Fabrication process of planar membrane hand. Reproduced with permission [35]. Copyright 2022, IEEE.

distance between the two membrane fingers, the soft membrane needs to be inflated by 20 mm in order to grasp objects of various sizes. Therefore, the target displacement of the inflated membrane was set to 20 mm.

The displacement of the membrane center was measured using a micro-laser ranging sensor (HGC1050-P, Panasonic Corporation, Japan), which was positioned toward the membrane finger and pointed at the center of the finger, as shown in Figure 3.3. The measurements were conducted by gradually increasing the applied pressure in 1 kPa intervals, starting from 8 kPa, until the membrane displacement reached 20 mm. Experimental snapshots with different air pressures are shown in Figure 3.3.

The results indicated that the fingers reached a 20 mm displacement at 16 kPa and 21 kPa for the membranes fabricated with Dragon Skin 20 and Dragon Skin 30, respectively. To avoid damaging objects during grasping,

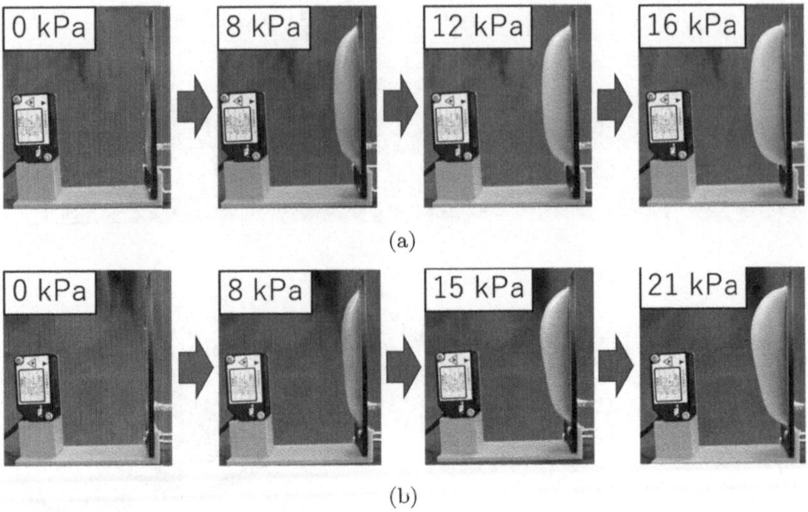

FIGURE 3.3
Inflation tests on fabricated fingers using (a) DS20. (b) DS30. Reproduced with permission [35]. Copyright 2022, IEEE.

planar membrane fingers fabricated with Dragon Skin 20 were selected for the subsequent grasping tests.

3.1.4 Hand prototype for grasping multiple cucumbers

The proposed planar membrane hand was used to package multiple cucumbers simultaneously. To maximize space efficiency and fit multiple cucumbers into the package, it is essential to minimize the gaps between neighboring cucumbers during packaging. At the grasping position, the cucumbers are aligned using a wave conveyor, creating enough space for the gripper to enter. After grasping, the gripper should be able to reduce the gaps before packaging.

To achieve this, a grasping strategy was proposed, as shown in Figure 3.4a. Each cucumber was grasped by two membrane fingers, with the grasping positions shifted between neighboring cucumbers. This strategy allowed the gap between neighboring cucumbers to be reduced to the thickness of a single membrane finger, approximately 3.0 mm. Based on this strategy, a robotic hand, as shown in Figure 3.4b, was assembled. A parallel link mechanism was used to shrink the gaps, with a pneumatic cylinder driving the mechanism.

The prototype of the planar membrane hand is shown in Figure 3.5a, and the experimental system used for packaging five cucumbers is depicted in Figure 3.5b. The membrane hand was mounted onto a robot arm (HSR065A1, DENSO WAVE Inc., Japan), which was programmed to perform pick-and-place motions. A wave conveyor was employed to align the cucumbers before grasping. Experimental snapshots are shown in Figure 3.6.

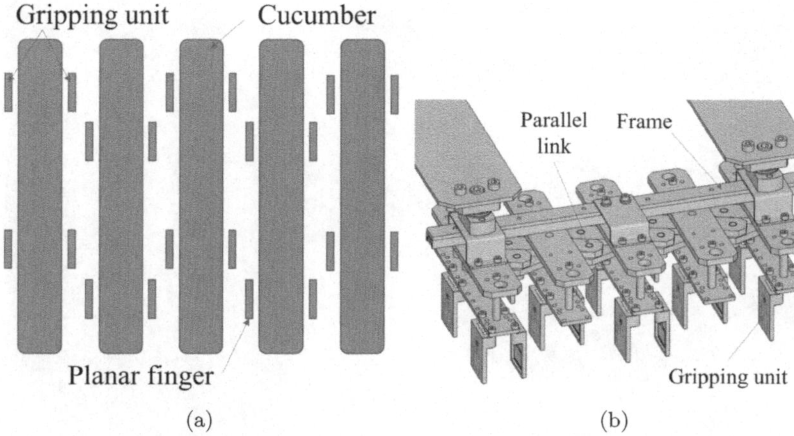

FIGURE 3.4
Finger configuration (a) and hand assembly (b). Reproduced with permission [31]. Copyright 2020, IEEE.

Upon grasping, the total distance between the five cucumbers held by the hand was approximately 235 mm (Figure 3.6a). After shrinking the gaps, the distance reduced to 165 mm (Figure 3.6b). The interval between neighboring cucumbers was minimized to the thickness of the membrane finger. The appearance after placing the cucumbers in the package is shown in Figure 3.6c.

The same pick-and-place experiment was conducted ten times with different cucumbers, transferring a total of 50 cucumbers. Only one cucumber dropped during the experiments, due to its large curvature, which made it

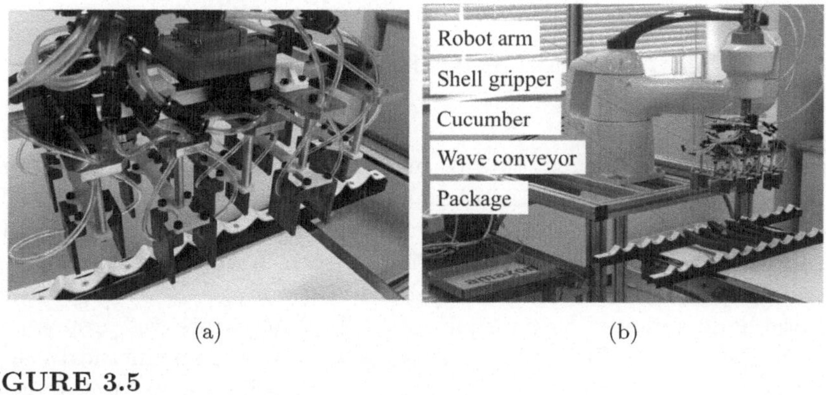

FIGURE 3.5
Manufactured hand prototype (a) and experimental setup (b). Reproduced with permission [31]. Copyright 2020, IEEE.

FIGURE 3.6
Experimental snapshots of the packaging test: (a) the cucumber grasping, (b) the gap shrinking, and (c) the packaged cucumber. Reproduced with permission [31]. Copyright 2020, IEEE.

difficult for the hand to grasp it stably. The takt time for each pick-and-place test was approximately 6.24 s, which is below the required 7 s. These packaging experiments demonstrated the capability of membrane hand to handle multiple cucumbers simultaneously.

3.2 Bellows Actuated Parallel Hand

3.2.1 Concept

When considering grasping objects in unstructured environments or bin-picking scenarios, the robotic hand may unintentionally contact surrounding objects or the environment. This could potentially lead to damage to either the robotic hand itself or the neighboring objects. To mitigate such damage, incorporating softness into the robotic hand is advantageous. To achieve this, a robotic hand actuated by multiple soft bellows actuators was proposed, as shown in Figure 3.7a. The bellows actuators are soft and pneumatically actuated, enabling an open-and-close motion. Two rigid fingers are used to grasp the object between them. Since the fingers and flange are connected by soft bellows actuators, any impact from contact is absorbed by the actuators, thus preventing damage to both the environment and the robotic hand.

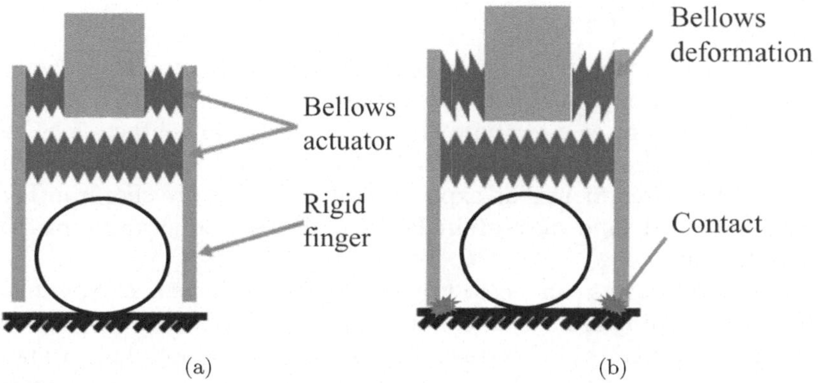

FIGURE 3.7
Concept of bellows actuated hand (a) and impact induced by contact with environment (b).

3.2.2 Hand prototype

The bellows actuators were 3D printed using TPU material, a flexible material with a Shore hardness of A85-A95. This material can be easily fabricated using an FDM 3D printer with the appropriate nozzle temperature settings. The rigid fingers and flange were 3D printed using PLA material. The prototype of the hand is shown in Figure 3.8a in its initial state. By applying air pressure to the long bellows actuators, the robotic hand opens, as shown in Figure 3.8b. Conversely, by applying vacuum pressure to all the bellows actuators, the hand closes to perform a grasp, as shown in Figure 3.8c. To increase the friction coefficient for stable grasping, a soft rubber sheet cushion was attached to the rigid fingers.

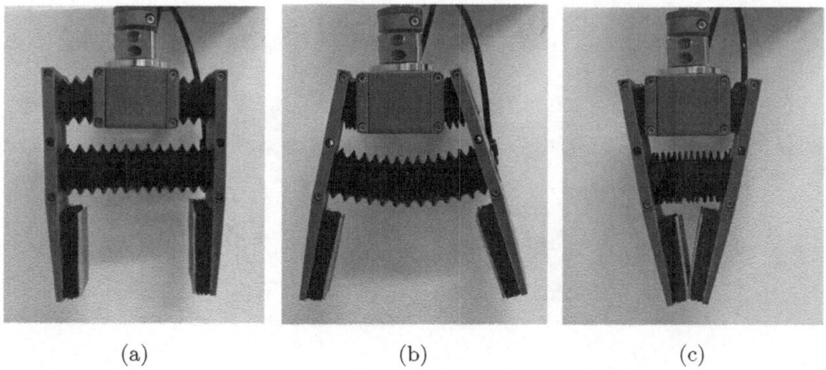

FIGURE 3.8
(a) Bellows actuated hand prototype and its (b) open and (c) close states.

3.2.3 Experiments

The robotic hand was tested on grasping various agricultural products, including white radish, potato, carrot, and burdock root. The first test involved a normal pick-and-place operation without any external contact. A SCARA robot (RH-CH series, Mitsubishi Electric Corp., Japan) was used to generate the pick-and-place motion. The experimental snapshots are shown in Figure 3.9. The developed hand successfully handled all the products at a maximum speed of 1200 mm/s.

The second test was conducted using the same products and experimental setup, but with programmed intentional contacts during the robot motion. Three contact scenarios were tested: (1) contact between both fingers and the ground (see Figure 3.10a), (2) contact between one finger and the environment (see Figure 3.10b), and (3) contact between one finger and a neighboring product (see Figure 3.10c). These scenarios are typical in agricultural applica-

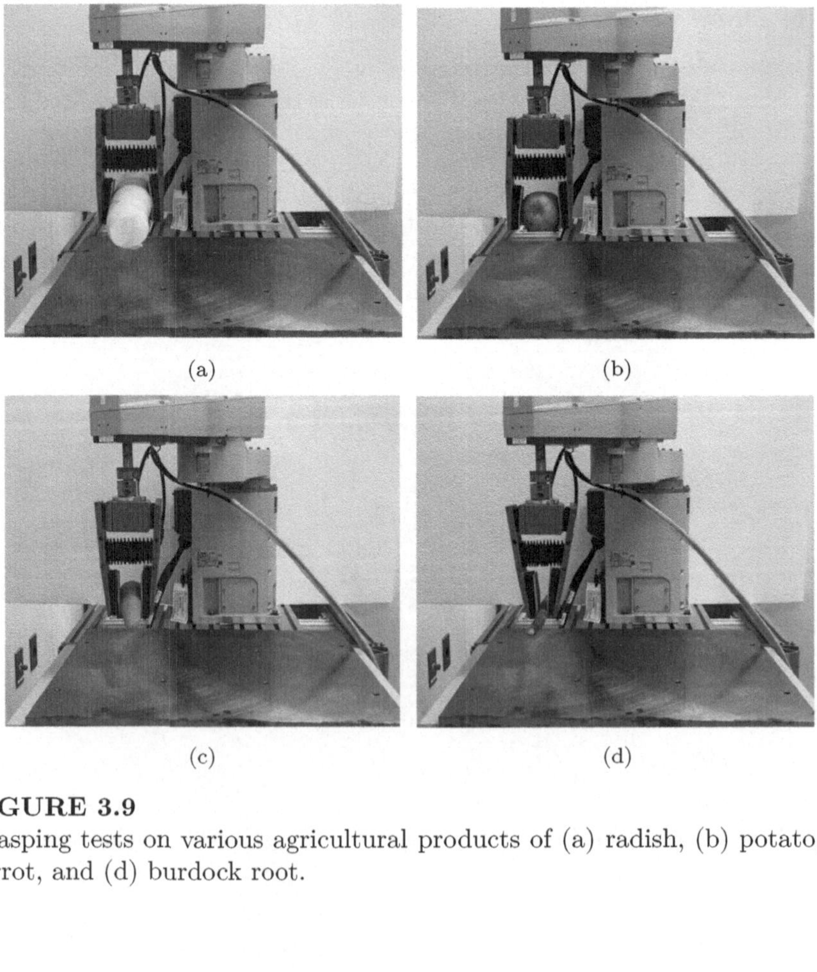

(a) (b) (c) (d)

FIGURE 3.9
Grasping tests on various agricultural products of (a) radish, (b) potato, (c) carrot, and (d) burdock root.

(a) (b) (c)

FIGURE 3.10
Grasping tests with various kinds of contacts with (a) floor, (b) step, and (c) neighboring product.

tions, where unstructured environments are common. Experimental snapshots in Figure 3.10 show that the proposed robotic hand successfully handled agricultural products in these contact scenarios.

4

Robotic Hands for Kitchen and Restaurant

This chapter introduces soft robotic hands tailored for kitchen and restaurant automation, with applications in food presentation, dish collection, and dish loading into washing machines. It covers the desired operational tasks, design principles, material selection, fabrication processes, and experimental evaluation of each soft robotic hand. Four robotic hands are presented: two are designed specifically for dishwashing applications, while the other two are developed for arranging tempura.

4.1 Dishwashing Bellows Hand

In the food service industry, particularly in restaurants and meal supply centers for schools and hotels, dishwashing machines are widely used. However, human workers are still needed to load and unload dishes. Loading is especially challenging, as plates are often stacked in piles and coated with water and oil after prewashing. While humans can easily grasp and remove plates from these piles (see Figure 4.1a), it remains difficult for a robotic hand to perform this task quickly enough to meet efficiency standards. Plates are typically heavy (around 500 g), and the water and oil reduce the friction coefficient, making them slippery to handle.

Kosuge *et al.* developed a robotic system featuring a hook-structured hand designed to automate dishwashing tasks [36,37]. The hook-structured hand includes an active finger and a passive hook. During operation, the hook inserts into the gap between stacked plates, and the active finger moves to secure the plate. Supported by a visual feedback system [38], this robotic setup successfully executed loading and unloading tasks. However, to avoid strong collisions between the plate and the rigid robotic hand, the grasping operation had to be performed at a reduced speed.

4.1.1 Hand concept

In [39], a robotic hand powered by soft bellows actuators was proposed (see Figure 4.1b). In the dish-loading process, pre-washed plates are usually wet, and their orientation must be adjusted from horizontal to inclined before

DOI: 10.1201/9781003372431-4

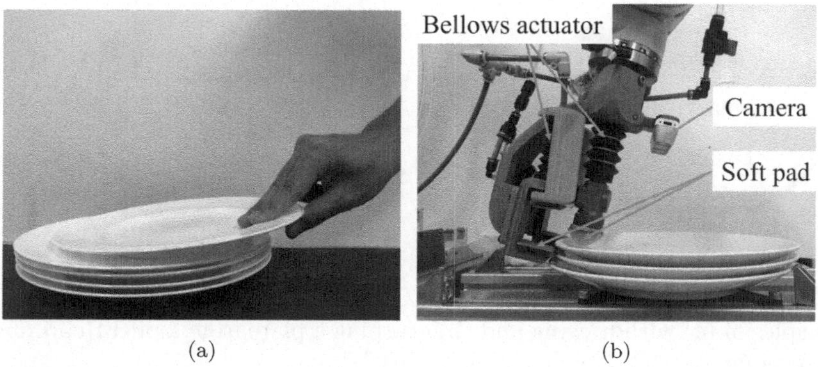

FIGURE 4.1
Withdrawing and grasping a plate from a pile using (a) human hand and (b) the developed robotic hand. Reproduced with permission [39]. Copyright 2021, IEEE.

placing them into the rack. Suction pads were avoided for this application due to maintenance challenges and the risk of handling failures.

Two main reasons motivated the choice of bellows actuators over electric motors. First, the bellows actuator provides compliance, allowing it to adjust flexibly upon contact with the dish plate. This flexibility enables higher operational speeds without concerns about potential collisions. Second, bellows actuators are easy to maintain in high-humidity environments, as they can be made waterproof and are suitable for water cleaning and steam sterilization.

The robotic hand is capable of both retrieving and securing plates. To improve grip, specially designed soft pads were added to the contact surfaces, ensuring stable friction between the hand and the plate, even when the plate is covered in water and oil.

The plate handling task consists of two main operations: withdrawing and grasping. The withdrawing operation involves generating a tangential motion to pull out the top plate from a stack, while the grasping operation holds the plate securely. To replicate human-like plate withdrawal, a mechanism was developed, as illustrated in Figure 4.2a. This design incorporates two bellows actuators arranged perpendicularly. The vertical actuator produces a compressive force, and the horizontal actuator performs the tangential withdrawing motion.

Following the withdrawal, a clamp-like grasping mechanism was designed to secure the plate, as shown in Figure 4.2b. This mechanism comprises three components: a U-shaped fixed jaw (shown in purple) attached to the robot arm for stable support during grasping, and two vacuum-driven bellows actuators that connect the fixed jaw to a movable jaw (shown in green) to perform the grasping motion.

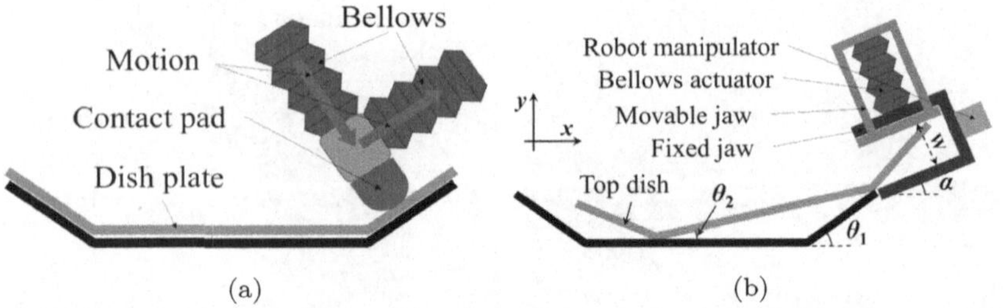

FIGURE 4.2
Principles of (a) withdrawing and (b) grasping a plate from a pile. Reproduced with permission [39]. Copyright 2021, IEEE.

4.1.2 Hand design

As shown in Figure 4.3, the robotic hand comprises a single withdrawing mechanism (Figure 4.3b) and two grasping mechanisms (Figure 4.3c). This design incorporates four bellows actuators in total—two for the withdrawing function and two on either side of the withdrawing mechanism for grasping. For the withdrawing operation, positive pressure is applied to the vertical actuator, while vacuum pressure is applied to the horizontal actuator. A half-cylinder-shaped soft pad at the tip of the mechanism enhances friction.

The grasping mechanism, designed with a clamp-like structure, consists of a fixed jaw (bottom) and a movable jaw (top). When vacuum pressure is applied, the two jaws close, securing the object. Flat friction pads are attached to the surfaces of both jaws to improve grip. Apart from the bellows actuators, all components were 3D-printed with PLA material. The bellows actuators were fabricated using a Prusa I3 MK3 3D printer (Prusa Research, Prague) with NinjaFlex material (NinjaTek, Fenner Drives, PA). The robotic hand measures 204 mm × 141 mm × 140 mm, with a center distance of 92 mm between the two grasping actuators and a total weight of 755 g.

4.1.3 Experiments

The experimental setup (Figure 4.4a) includes a robot manipulator (RV-5AS-D, Mitsubishi Electric Corp., Tokyo), the developed robotic hand, and an Intel RealSense D435 camera. For the withdrawing operation, pressures of 30 kPa and −30 kPa were applied to the vertical and horizontal bellows actuators, respectively. The grasping operation used a vacuum pressure of approximately −50 kPa. The camera provides positional and depth data for the top plate.

A pick-and-place motion was programmed for the robot manipulator. At the pick position, the robotic hand withdraws and grasps the plate based on the top plate's detected position and depth. After grasping, the robot moves to a predefined posture to insert the plate into the rack.

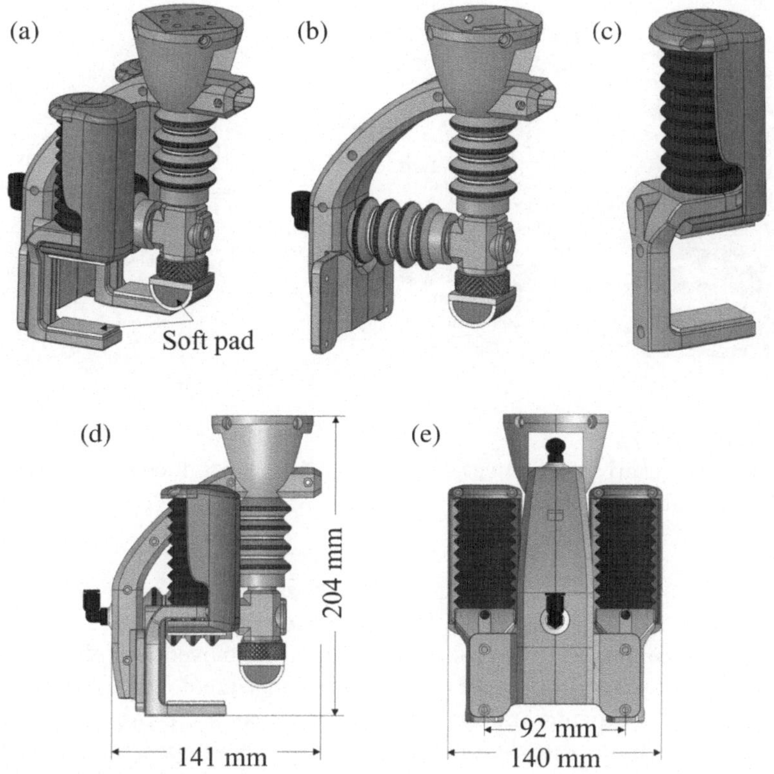

FIGURE 4.3
3D CAD drawings of the dish withdrawing hand: (a) the isometric drawing of the hand, (b) the withdrawing mechanism, (c) the grasping mechanism, (d) the side view, and (e) the back view. Reproduced with permission [39]. Copyright 2021, IEEE.

Experiments were conducted on three types of plates of varying sizes and weights (Figure 4.5a). Plate types 1 and 2 are plastic, with type 1 having the steepest lip angle at 48.6°. Plate type 3 is ceramic and weighs 515 g. During each experiment, the camera first identified the plate (Figure 4.5b). The plate's center was calculated by fitting a circle to its boundary, and the contact point was predefined based on the distance from the center to the edge. The robotic hand's current position is shown as the dot in between the contact point and the plate center. The camera's depth data was used to calculate the distance between the contact point and the robotic hand. The robotic hand was then moved to the contact point using a visual feedback algorithm, aligning the two points (Figure 4.5c).

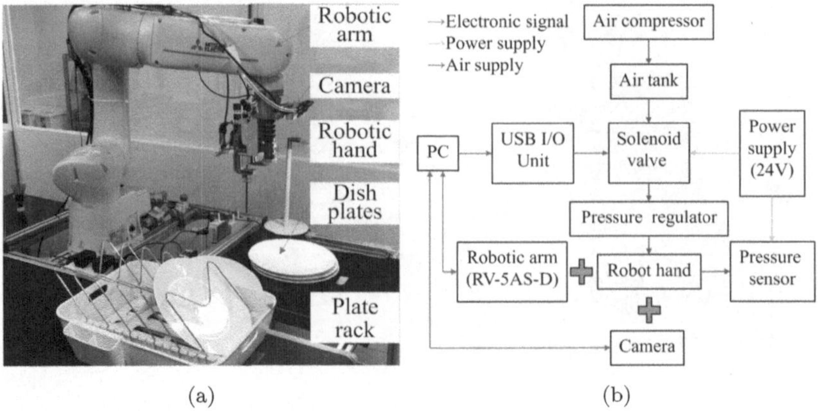

FIGURE 4.4
Experimental setup (a) and system diagram (b). Reproduced with permission [39]. Copyright 2021, IEEE.

Following plate recognition at the initial position (Figure 4.6a), the robotic hand was lowered to withdraw a plate from the pile and execute the grasping operation. Once grasped, the robotic manipulator moved the plate to a designated rack position and placed it securely. Experimental snapshots showing the steps of withdrawing and grasping various plates are provided in Figure 4.6b–f. These experiments demonstrated that the developed robotic hand can successfully extract and grasp plates from a stack and transfer them to a set location. The time required to handle each plate is approximately 36 s.

4.2 Dishwashing Hybrid Hand

In the food service industry, robots have been adopted for automating food service tasks. Tzou *et al.* developed a mobile restaurant robot equipped with a laser positioning system for rapid and precise positioning and guidance [40]. Yu *et al.* designed a restaurant robot capable of taking orders, fetching and delivering food, and handling payment for customers [41, 42]. Guo *et al.* introduced a service robot system utilizing dual vision positioning, combining global monitoring with mobile depth detection [43]. Shimmura *et al.* implemented an automated guided vehicle system in a Japanese cuisine restaurant, examining its impact on reducing work hours and enhancing labor productivity and service quality [44].

On the other hand, robots are rarely utilized in the back areas of restaurants, hotels, hospitals, and similar settings. Most operational tasks, such as cooking and dishwashing, still rely heavily on human labor. In [45],

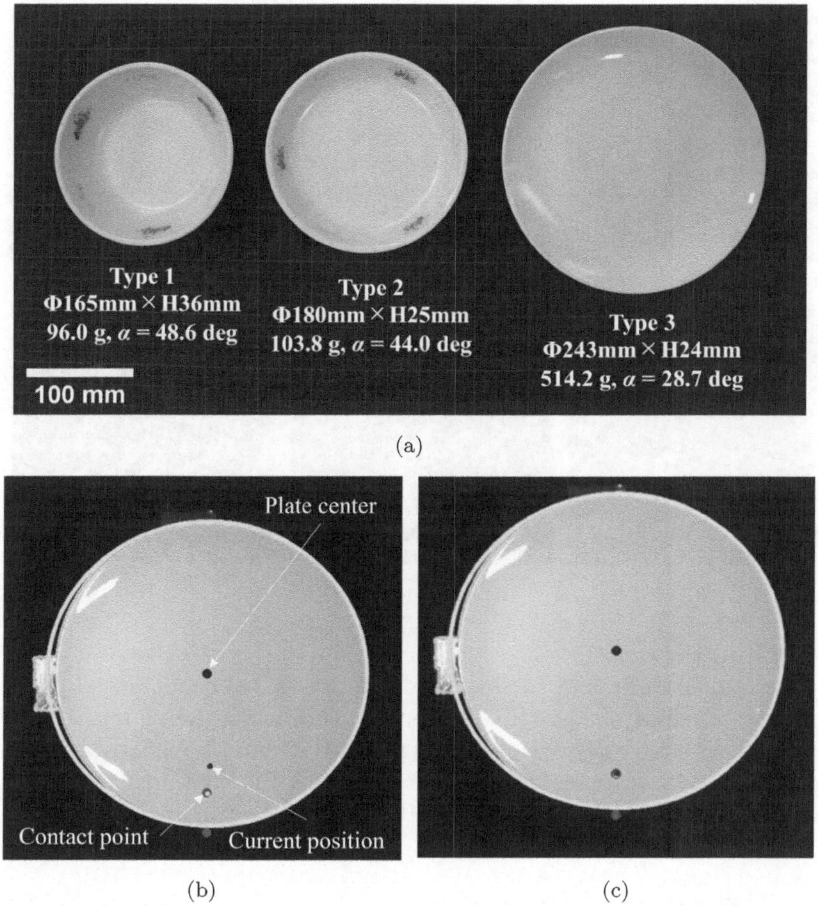

FIGURE 4.5
Plates used in experiments (a) and plate recognitions with (b) initial contact position and (c) final matched position using visual feedback. Reproduced with permission [39]. Copyright 2021, IEEE.

dishwashing automation was explored. As illustrated in Figure 4.7, dishwashing tasks can be primarily divided into putting-in and taking-out operations, which can be handled by robots A and B, respectively. In the previous section, we discussed a robotic system with a 6-DOF manipulator, a 3D camera, and a bellows-driven robotic hand designed for the putting-in operation. This system could effectively pick and securely grasp dish plates, even when covered with water or oil. However, challenges arose in performing the taking-out operation due to differences in requirements between the two tasks. Therefore, this section introduces a robotic system specifically designed for the taking-out operation.

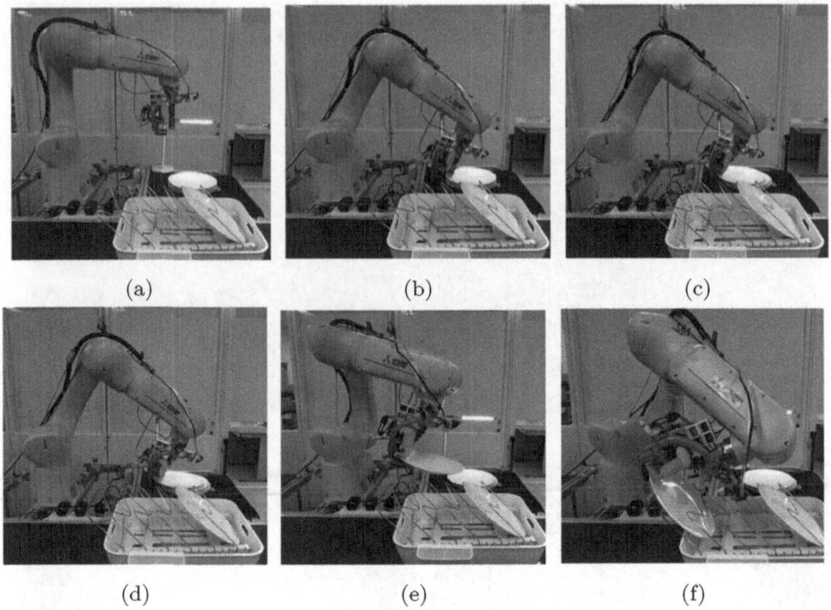

FIGURE 4.6
Experimental snapshots of tableware loading tests: (a) robot initial position, (b) gripper touching the plate, (c) gripper withdrawing the plate, (d) gripper grasping the plate, (e) gripper holding the plate and translating it to the container, and (f) gripper loading the plate into the container. Reproduced with permission [39]. Copyright 2021, IEEE.

4.2.1 Hybrid hand concept

An actual scenario of tableware after dishwashing is shown in Figure 4.8. The tableware is randomly arranged in the rack, each with arbitrary orientations, making it a challenging task for a robotic system to recognize and handle them effectively. To simplify the handling process, the orientations of the tableware were categorized into two groups: horizontal and vertical orientations. For tableware in the horizontal orientation, suction cups are ideal as the flat surfaces provide a sufficient area for suction. In contrast, for tableware in the vertical orientation, parallel grippers are more suitable for grasping. To accommodate both orientations, a robotic hand capable of both suction and grasping was proposed.

The design and prototype of the hybrid robotic hand are shown in Figure 4.9. A pneumatic parallel gripper provides the grasping motion with two fingers, each equipped with four soft pads to securely grasp the tableware. Additionally, four pneumatic cylinders are mounted on the back of fingers, each with a suction cup at the end. During suction operations, the cylinders extend, bringing the suction cups into contact with the tableware. When not

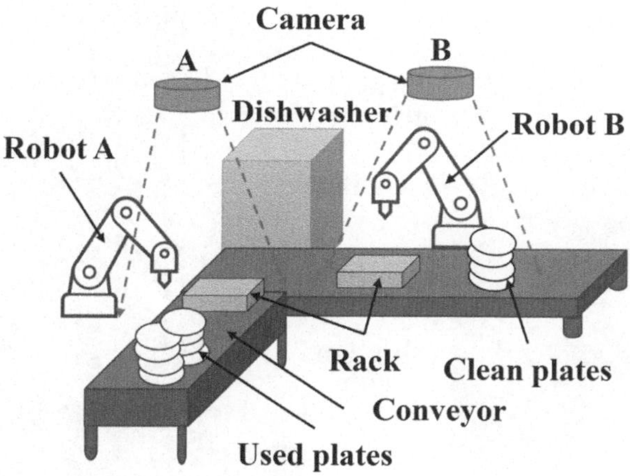

FIGURE 4.7
Concept scenario of robotic automation system for dishwashing. Reproduced with permission [45]. Copyright 2021, IEEE.

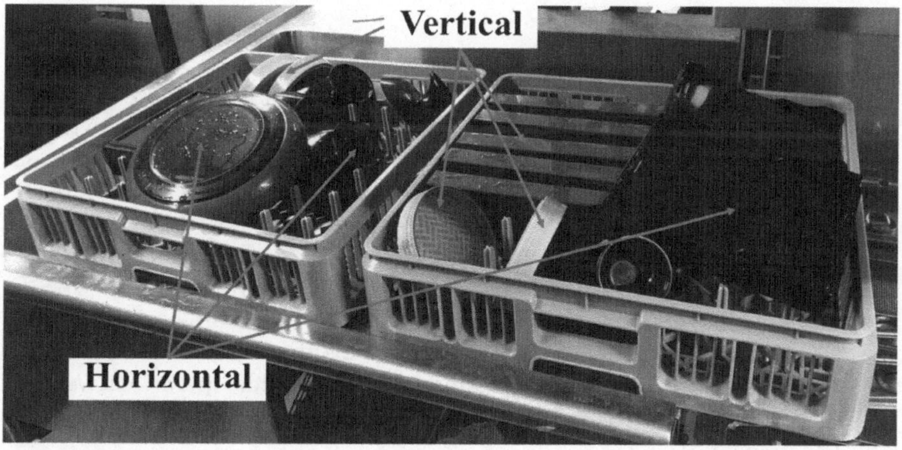

FIGURE 4.8
Actual scenario of tableware after dishwashing. Reproduced with permission [45]. Copyright 2021, IEEE.

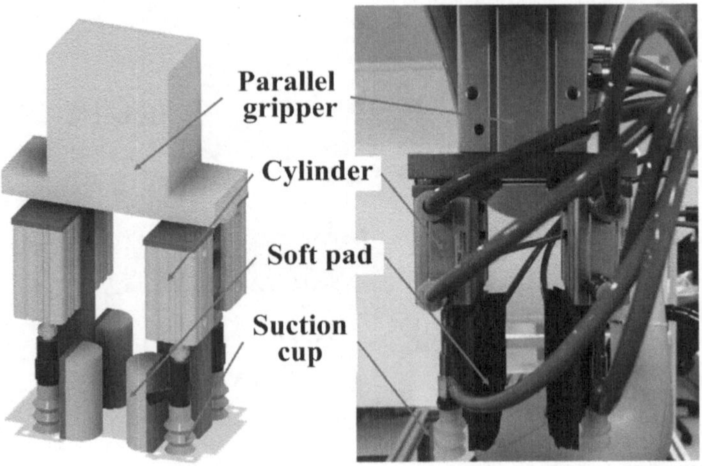

FIGURE 4.9
Design and prototype of the hybrid dishwashing hand. Reproduced with permission [45]. Copyright 2021, IEEE.

in use, the cylinders retract, keeping the suction cups clear to avoid interference with other movements. The robotic hand weighs approximately 1250 g and measures about 65 mm × 110 mm × 210 mm.

4.2.2 Experimental tests

The tableware items used in experiments are shown in Figure 4.10a. The salad bowls are ceramic, while the other tableware items are made of plastic. To simplify recognition and grasping tasks, we assume two distinct orientations for the tableware after dishwashing: vertical and horizontal, as depicted in Figures 4.10b and c.

Experiments were conducted using the setup shown in Figure 4.11a. The robotic hand was mounted onto a 6-DOF robotic manipulator (RV-5AS-D-S99, Mitsubishi Electric Corp., Tokyo). A 3D camera (RealSense D435, Intel Corp., CA) provided image data for object recognition. Five tableware items were randomly placed in the dishwashing rack. For suction operations, vacuum pressure of −60 kPa was generated using air ejectors (VUH07-66A, PISCO Co., Ltd., Nagano).

The operation flow for the experiments is shown in Figure 4.11b. First, the robot manipulator moved to its initial position, where an image of the tableware in the rack was captured. Recognition was then conducted to calculate the positions and orientations of the tableware items. Based on recognition scores, a handling target was selected, and either a grasping or suction

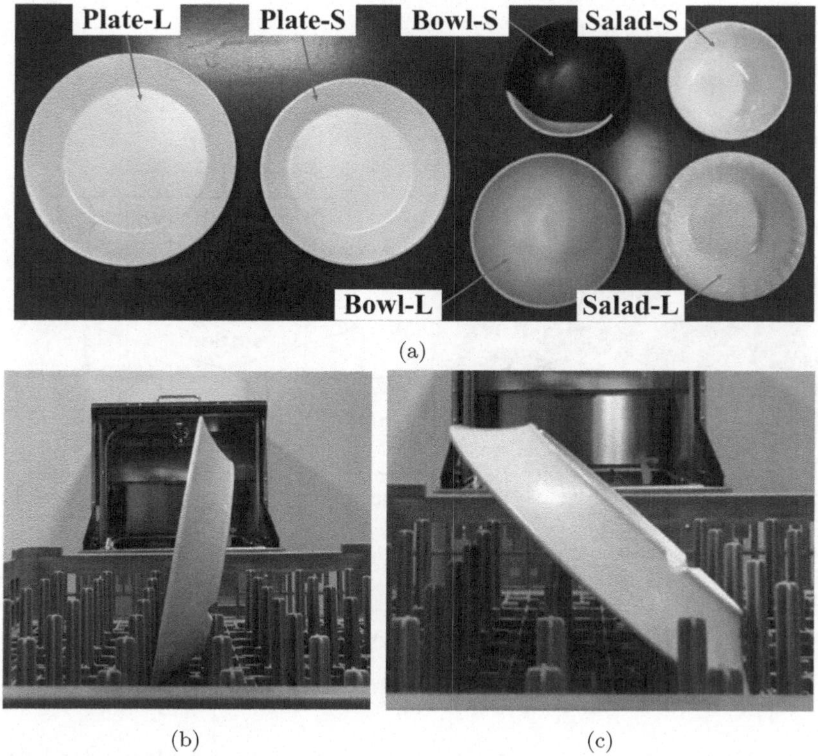

FIGURE 4.10
Tableware used in the tests (a) and the tableware orientations of vertical (b) and horizontal (c). Reproduced with permission [45]. Copyright 2021, IEEE.

operation was chosen depending on the tableware's orientation. The robotic hand then performed the grasping and translational movements to place the tableware at the designated position. The robot manipulator returned to its initial position after each operation, repeating the sequence until no target tableware was recognized in the rack.

Experimental snapshots of grasping the plate-L are shown in Figure 4.12. Following tableware recognition and grasping position and posture determination, the robotic hand was positioned to the required posture (Figure 4.12a). Next, it moved to the target grasping position to secure the tableware (Figure 4.12b). After grasping, the robot lifted the tableware to a set height (Figure 4.12c), then transferred it to a designated position (Figure 4.12d) and placed it there (Figure 4.12e). Finally, the robot returned to its initial position for the next operation (Figure 4.12f). The cycle time for one grasping operation was approximately 17 s.

Snapshots of handling plate-L and bowl-L using suction cups are shown in Figure 4.13. For plate-L, which has a sufficiently large surface area, four

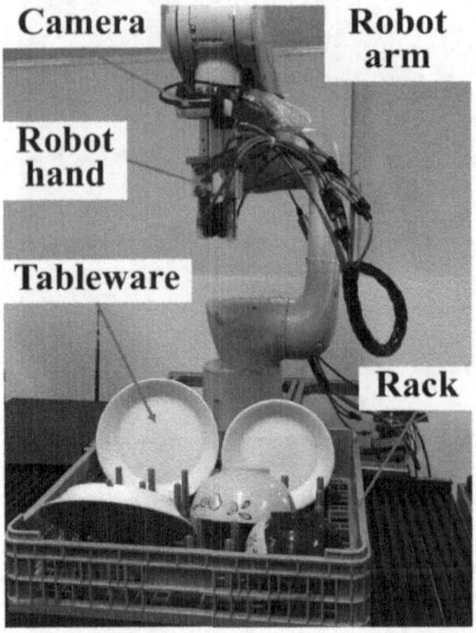

(a)

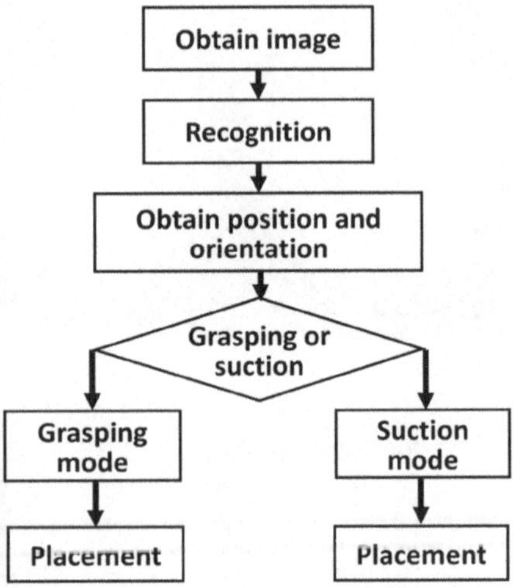

(b)

FIGURE 4.11
Experimental setup (a) and operation flow (b). Reproduced with permission [45]. Copyright 2021, IEEE.

suction cups were used to achieve a secure hold (Figures 4.13a–d. In contrast, only two suction cups were needed for handling bowl-L due to the small area available on the bowl's bottom surface (Figures 4.13e and f). The suction operation had a takt time of approximately 14 s, which was faster than the grasping operation due to the simpler placement motion enabled by suction. The total takt time for handling all five tablewares was approximately 140 s.

4.3 Three-Finger Hand for Tempura Arrangement

Tempura is a traditional Japanese dish consisting of battered and deep-fried seafood, meat, or vegetables. When serving tempura, it is often necessary to arrange various types of tempura in an aesthetically pleasing and predefined pattern on a dish, as shown in Figure 4.14a. Currently, this tempura arrangement task is performed by human laborers in the kitchen of restaurants, as depicted in Figure 4.14b. However, there is a growing demand for automation in this process.

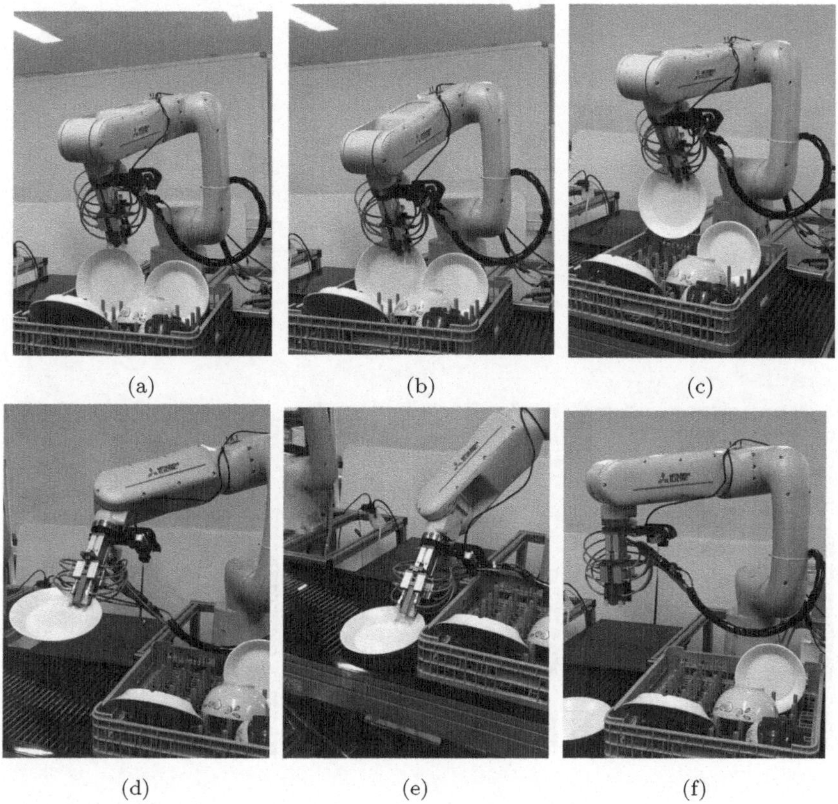

FIGURE 4.12
Experimental snapshots of grasping operations: (a) approaching to plate-L with the determined posture, (b) grasping plate-L at the determined position, (c) lifting plate-L to a predetermined height, (d) translating plate-L to the desired position, (e) placing plate-L at the desired position, and (f) returning to the initial position for the next grasp. Reproduced with permission [45]. Copyright 2021, IEEE.

4.3.1 Three-finger hand

To perform tempura arrangement, the robotic hand must meet several key requirements: (1) it should be adaptable to various types of tempura with different shapes and properties; (2) it needs to generate a small grasping force to avoid damaging the delicate tempura; and (3) it should have a simple structure, fabricated from metal materials to ensure low cost and ease of maintenance. To meet these criteria, a three-finger hand was proposed, as shown in Figure 4.15a. This hand consists of three rigid stick-like fingers and a parallel gripper. The fingers are arranged in an equilateral triangle configuration to grasp objects placed between them. To ensure three contact points with

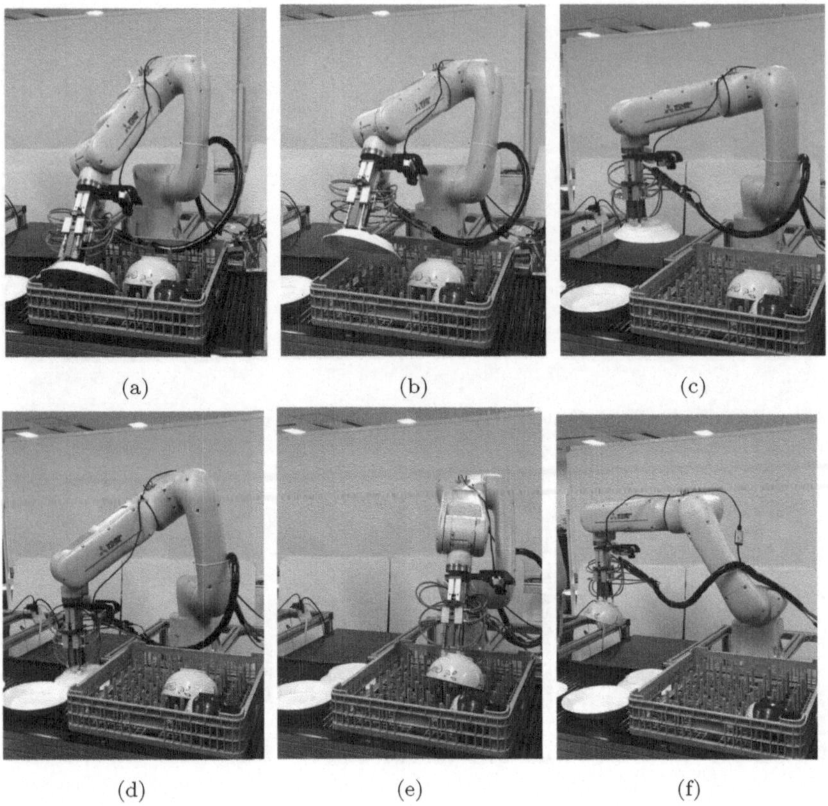

FIGURE 4.13
Experimental snapshots of suction operations: (a) approaching to plate-L with the determined posture, (b) holding plate-L using suction cups, (c) translating plate-L to the desired position, (d) placing plate-L at the desired position, (e) holding bowl-L using suction cups, and (f) translating bowl-L to the desired position. Reproduced with permission [45]. Copyright 2021, IEEE.

the object, a passive joint is used on one side of the two fingers, as shown in the middle of Figure 4.15a. When the parallel gripper closes, the fingers press against the object, and the two-finger side rotates to make contact with the object's surface, as illustrated in Figure 4.15b. Additionally, to achieve a small grasping force without the need for costly force sensors, a servo motor (XM430-W210-R, ROBOTIS) was used. The motor's current was calibrated to generate a grasping force of approximately 1.2 N.

(a) (b)

FIGURE 4.14
Experimental setup (a) and operation flow (b).

4.3.2 Tempura arrangement tests

Tempura arrangement tests were conducted in a restaurant's backyard using a robotic manipulator (MELFA, Mitsubishi Electric Corp., Japan), a belt conveyor, and the developed robotic hand, as shown in Figure 4.16a. Five types of tempura were used in the tests: shrimp, shishito pepper, sliced eggplant, sliced pumpkin, and sliced lotus root. These tempuras were prepared in a stainless steel container and transported to a location near the robot. A 3D camera (D435i, RealSense, Intel) was used to recognize the tempura in its initial position (Figure 4.16b). After the tempuras were recognized, the robotic manipulator was controlled to move the hand to the grasping position and pick up the targeted tempura (Figure 4.16c). The manipulator then translated to the placing position (Figure 4.16d) and placed the tempura in the predetermined position and orientation (Figure 4.16e). This procedure was repeated to arrange the remaining tempuras in a predefined order (Figure 4.16f). The final arrangement of tempura is shown in Figure 4.16g as an example.

4.4 Multi-Wire Hand for Tempura Arrangement

In the previous section, a robotic hand with three stick-like fingers was introduced, which successfully grasped and arranged five types of tempuras. The servo motor was calibrated to generate a total grasping force of 1.2 N without using a force sensor. Assuming the grasping force is evenly distributed across

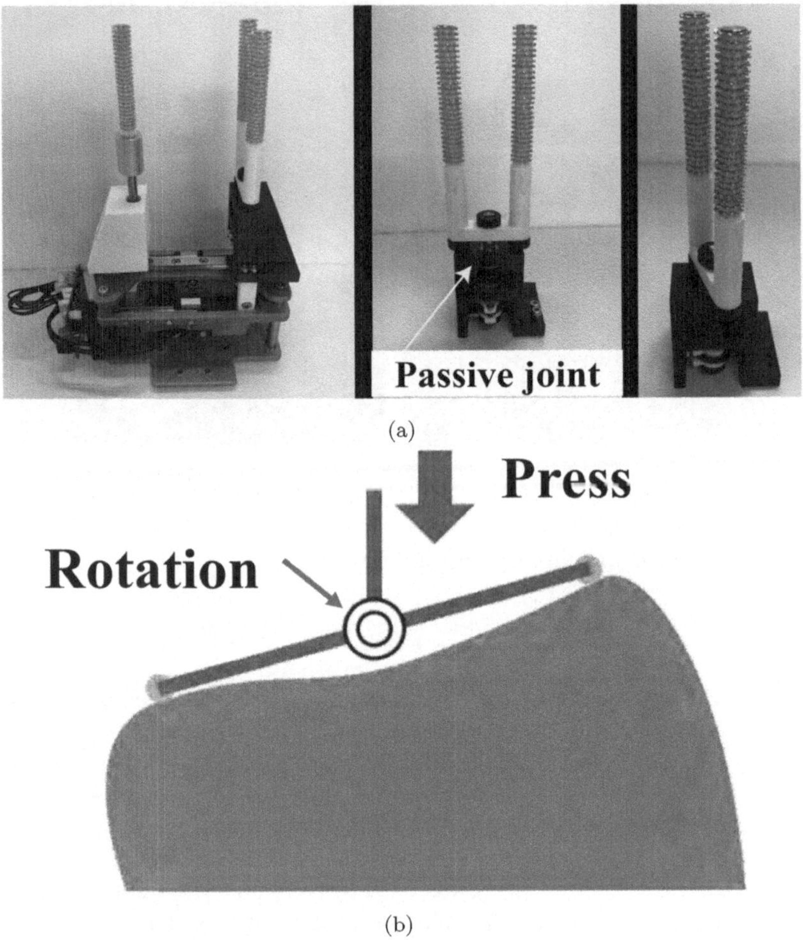

FIGURE 4.15
Hand prototype (a) and concept (b) of keeping three-point contact.

the three fingers, each finger exerts approximately 0.4 N force. Although this force may seem modest, it is relatively strong considering the fragile nature of tempura. Therefore, a robotic hand capable of producing a gentler grasping force was explored.

4.4.1 Multi-wire hand

As an extension of the three-finger hand, a multi-wire hand was proposed, as shown in Figure 4.17a. This design consists of a parallel gripper and two fingers, with each finger constructed from multiple thin wires. The number of wires can be customized based on specific requirements. In the prototype

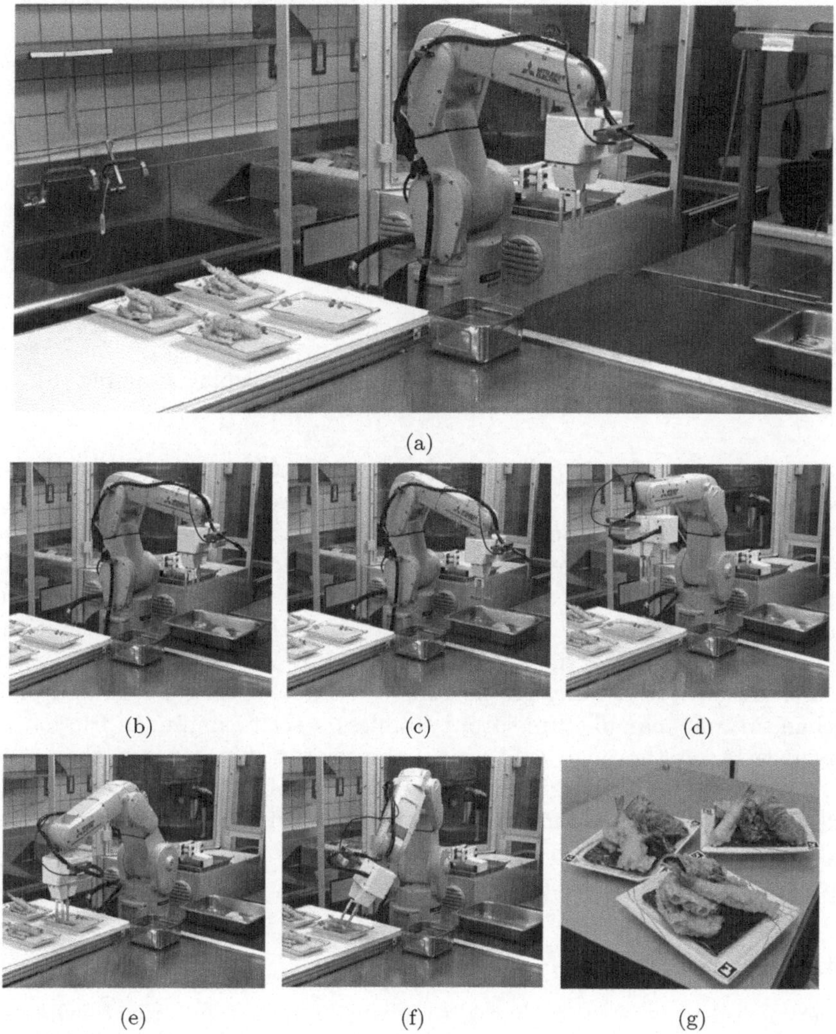

FIGURE 4.16
Experimental results on arranging five types of tempura: (a) experimental setup, (b) position for tempura recognition, (c) grasping of pumpkin, (d) position for placing pumpkin, (e) arrangement of pumpkin, (f) arrangement of fried shrimp, and (g) the final arrangement result.

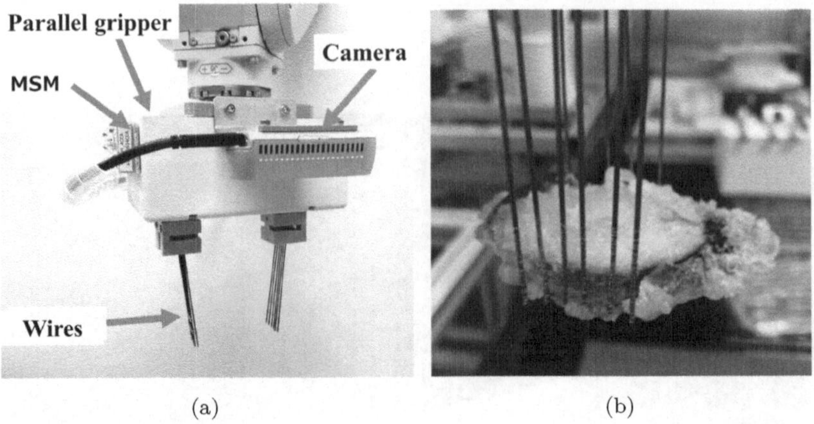

FIGURE 4.17
Prototype of the multiple wire hand (a) and food-grasping example (b).

shown in Figure 4.17a, six piano wires were used per finger, angled precisely so that they straighten when a grasping force of approximately 1.2 N is applied. With a total of 12 wires on the hand, each wire experiences a grasping force of around 0.1 N, which is gentle enough to avoid damaging delicate foods. The flexible and independent nature of each wire allows them to conform to the shape of the food, as shown in Figure 4.17b, ensuring an even distribution of the grasping force across all wires.

In addition, the parallel gripper was constructed using a rack-and-pinion mechanism and manufactured from stainless steel to minimize friction and resistance. After thorough calibration, the relationship between the applied current and the resulting grasping force was established, as shown in Figure 4.18. This calibration confirmed that the multi-wire hand could generate a grasping force of less than 0.5 N, making it well-suited for handling fragile items with minimal risk of damage.

4.4.2 Food arrangement tests

Experimental tests were conducted to arrange tempura in a rice bowl and an appetizer of steamed vegetables. The experimental setup is depicted in Figure 4.19. For the tempura rice bowl arrangement, the same robotic system used in Section 4.3 was employed. A rice bowl was prepared, and the robotic system arranged the tempura pieces within it. For arranging the steamed vegetable appetizer, a collaborative robotic manipulator (UR5e, Universal Robots, Denmark) was utilized. Experimental snapshots shown in Figure 4.20 demonstrate that the proposed multi-wire hand successfully handled both the tempura and steamed vegetables without causing any damage.

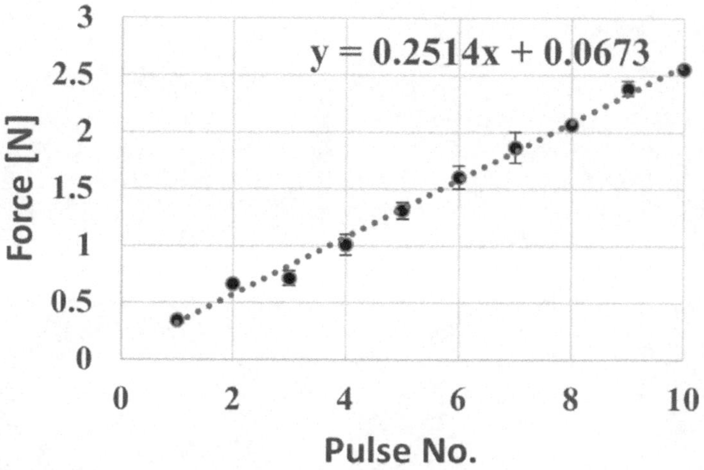

FIGURE 4.18
Force calibration results of multi-wire hand.

FIGURE 4.19
Experimental setup for grasping various food items.

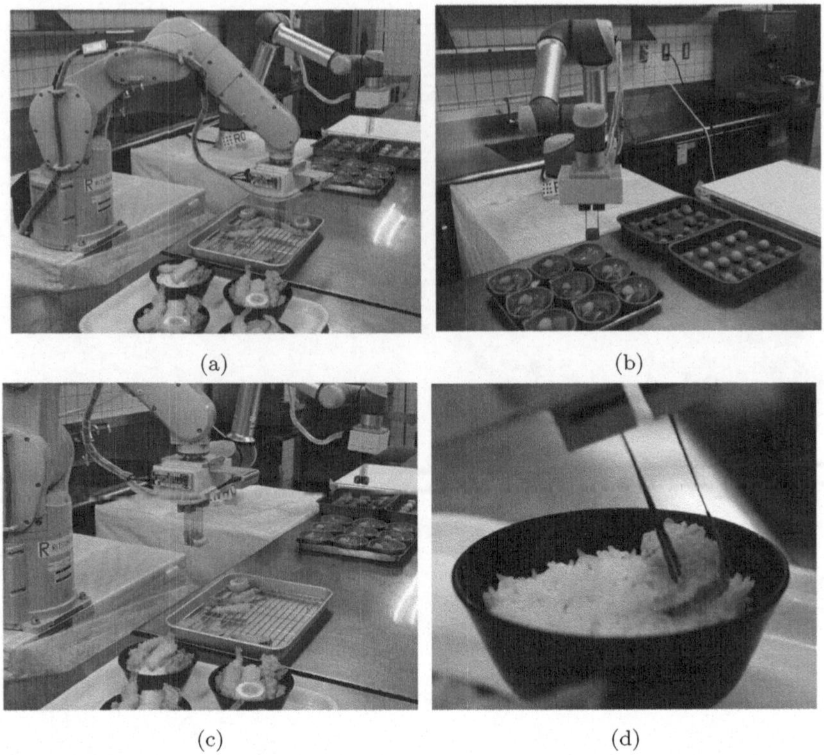

FIGURE 4.20
Grasping tests on various food items: (a) tempura shrimp, (b) steam boiled pumpkin, (c) tempura boiled egg, and (d) placement of tempura pumpkin.

5

Food Recognition

This chapter focuses on food recognition during manipulation. Recognizing the shapes, sizes, and locations of food items is crucial for effective manipulation. The chapter covers 2D/3D image acquisition and processing techniques, feature extraction methods, and learning-based recognition approaches.

5.1 Pattern Matching Based Approach

In actual food production lines, bin-picking scenarios, as shown in Figure 5.1a, are commonly encountered. While humans can easily perform such tasks at high speeds, it is a significant challenge for robotic systems. Before a robotic hand can grasp a target food item, the recognition of the food must first be completed, typically using vision-based methods. This section introduces and tests a straightforward approach based on pattern matching, utilizing the robotic system shown in Figure 8.3. The setup includes a SCARA robot (HSR-065, DENSO WAVE), a 3D camera (D435i, RealSense, Intel), and the planar shell hand, which has been introduced in Chapter 3.

5.1.1 Recognition principle

Recognition using template matching relies on 3D point cloud data from both the template point cloud and the measurement point cloud, as shown in Figure 5.2a. First, the normal vector is calculated from the point cloud data, and the feature value is determined. Next, global alignment is performed, as illustrated in Figure 5.2b. Finally, ICP (iterative closest point) alignment is carried out to match the two point clouds, as depicted in Figure 5.2c.

5.1.2 Recognition tests

To test the pattern matching method, a 3D point cloud of deep-fried shrimp in a bin-picking scenario was captured using a 3D camera (D435i, RealSense, Intel), as shown in Figure 5.3a. A template point cloud, shown in Figure 5.3b, was used for pattern matching. The matched model is displayed in Figure 5.3c, with a cross indicating the position of the matched model, which is calculated

DOI: 10.1201/9781003372431-5

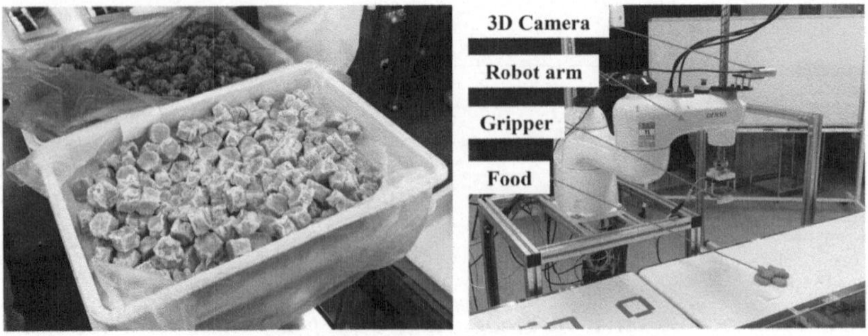

(a) (b)

FIGURE 5.1
Scenario of random food picking (a) and experimental setup (b).

(a) (b) (c)

FIGURE 5.2
Principle of 3D-template matching: (a) point clouds input, (b) global registration, and (c) ICP registration.

as the center of gravity of the template point cloud. The orientation of the detected fried shrimp was determined by calculating the moment of inertia of the template point cloud, as shown by the straight line in Figure 5.3d. With the detected position and orientation, the robotic manipulator and hand can be controlled to grasp the detected fried shrimp.

5.1.3 Grasping tests

Grasping tests with food recognition were conducted on four food items: croquette, spring roll, fried shrimp, and shumai, as shown in Figure 5.4. The recognized food items are displayed in the second row of Figure 5.4. After grasping the first target, pattern matching was performed again to recognize the second target. This procedure was repeated until no more targets matched the template. Experimental snapshots of grasping the croquette and spring roll are shown in Figure 5.5.

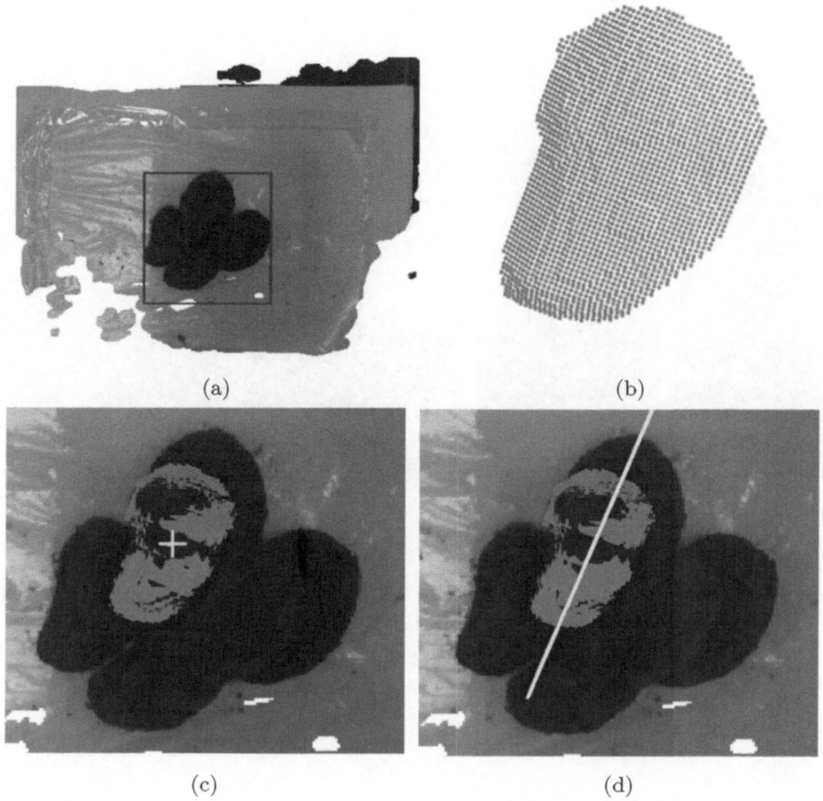

FIGURE 5.3
Recognition of deep-fried shrimp using pattern matching: (a) the measured point cloud, (b) the template point cloud, (c) the template matching and position extraction, and (d) posture extraction.

5.2 Image Processing Based Approach

Considering the scenario where food items are prepared or supplied separately on a flat surface, such as a table, tray, or belt conveyor, image processing can be employed to recognize food and determine the grasping positions. In this case, we take the example of automatic tempura arrangement to illustrate the details of how image processing can be used to obtain the grasping positions.

The tempura arrangement operation is commonly performed in the backyard of a restaurant. Several types of tempura are typically prepared in separate trays, and the chef arranges them on a plate in a predetermined order.

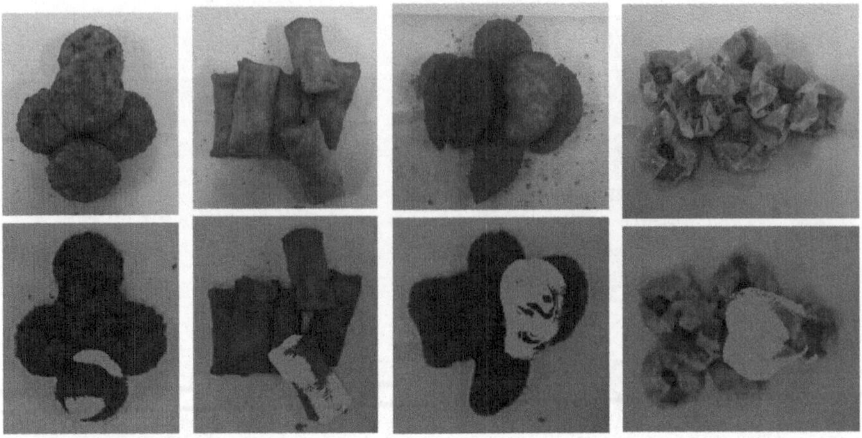

(a) (b) (c) (d)

FIGURE 5.4
Recognition of different food items for random-picking experiments: (a) the croquette, (b) the spring roll, (c) the deep fried shrimp, and (d) the shumai.

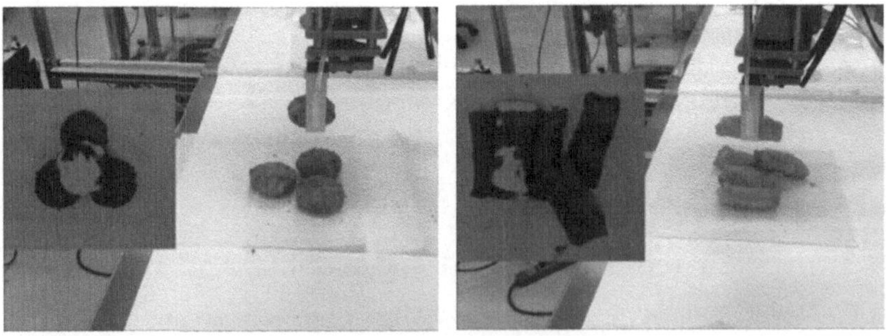

(a) (b)

FIGURE 5.5
Grasping experiments based on food recognition using pattern matching: (a) grasping croquette and (b) grasping spring roll.

(a) (b)

FIGURE 5.6
System setup (a) for automatic tempura arrangement and the arrangement target (b).

The goal is to present the tempura beautifully in a 3D arrangement. To automate this task, a robotic system was developed, as shown in Figure 5.6a. The system consists of an industrial robotic manipulator (MELFA RV-4FRL, Mitsubishi Electric Corp.), a 3D camera (RealSense D435i, Intel), and a parallel gripper. Five different types of tempura were prepared in separate trays. The camera was used to recognize each tempura, which was then grasped by the gripper one by one. The tempura were translated to the target plate and arranged according to the predetermined order and orientation. Examples of completed tempura arrangements are shown in Figure 5.6b.

An example of the obtained image is shown in Figure 5.7a. To accurately recognize the tempura against the background, four AR markers were placed at the corners of the tray, and the depth information of these markers was acquired. Using the depth data from the markers and the tray, the background behind the tempura could be removed, allowing the tempura to be extracted, as shown in Figure 5.7b. Next, the ellipse fitting function in OpenCV was used to approximate the boundaries of the tempura with ellipses. Examples of the ellipse fitting for shrimp, lotus root, and pepper are shown in Figures 5.7b–d, respectively. Once the ellipse fitting was completed, the position and orientation of the tempura were determined by the center and the long axis of the ellipse.

For recognizing the shrimp and sliced pumpkin, ellipse fitting alone is insufficient, as the tempura arrangement requires the shrimp's tail to be placed upward and the curvature of the sliced pumpkin to face inward. To determine the shrimp's tail, color information can be utilized, as the tail and body have significantly different colors. By setting a color threshold, the tail can be extracted from the shrimp, as shown in Figure 5.8a. To identify the curvature

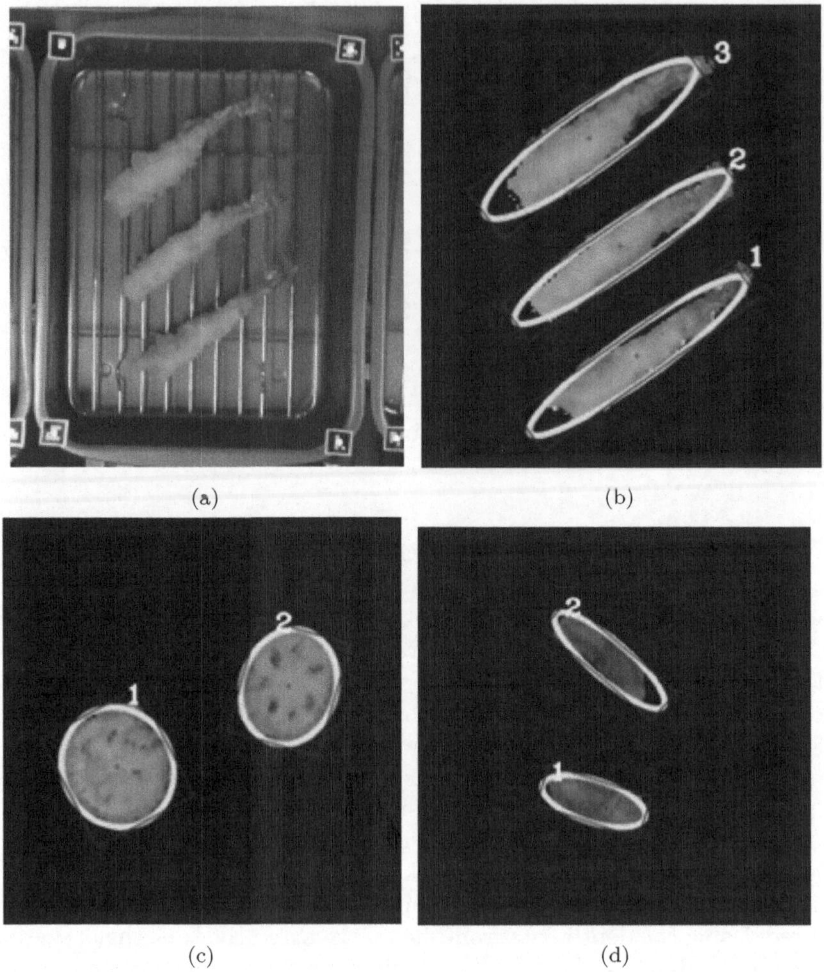

FIGURE 5.7
Tempura extraction and ellipse approximation: (a) tempura RGB image and ellipse approximations for (b) shrimp, (c) lotus, and (d) pepper.

direction of the sliced pumpkin, three points were selected from each side, and the curvatures of both sides were calculated, as shown in Figure 5.8b. By comparing the curvature values, the direction of the curvature can be determined. With this information on the shrimp's position, orientation, tail, and the pumpkin's curvature, the grasping and arrangement can be automated. The final arrangement of the tempura is shown in Figure 5.6b.

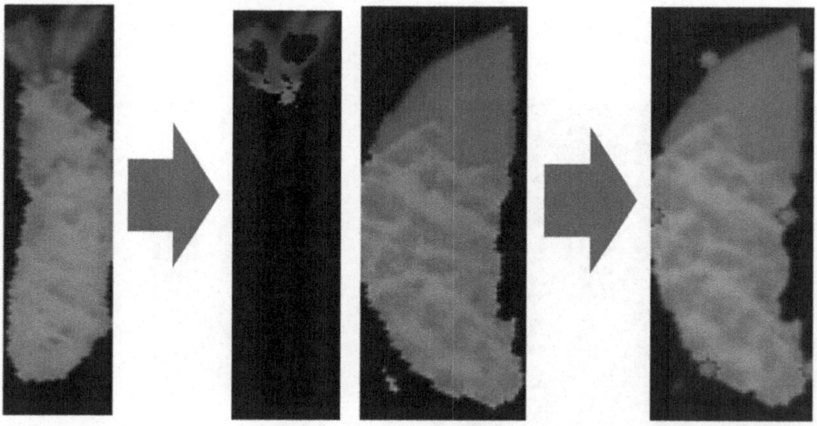

(a) (b)

FIGURE 5.8
Recognition of shrimp (a) and pumpkin (b) orientations.

5.3 Deep Learning Based Approach

Image processing can effectively extract the food target and identify its boundary, but it struggles to recognize the categories of different food items. To address this, the YOLO (you only look once) algorithm [46] is often used for object category recognition. In the context of tempura arrangement, YOLO can be employed to recognize different tempura categories, as shown in Figure 5.9, where different colored bounding boxes indicate various tempura categories. Once the bounding boxes are obtained, the position and orientation of each tempura can be determined using the image processing methods described earlier. This recognition process can be efficiently implemented by adding training data consisting of separate tempura images. However, real-world applications often present more complex scenarios, such as bin-picking situations (see Figure 5.1a), which can complicate the recognition process.

In bin-picking scenarios, food items are often randomly scattered and may overlap within a container. Recognizing each individual item in such scenarios becomes a challenging task due to the similar features shared by many food items. However, by training YOLO with a large dataset that includes various overlapping situations, it becomes possible to recognize food items even in complex bin-picking scenarios, as demonstrated in Figure 5.10 [47].

However, training an accurate and robust model requires a substantial amount of data, which is one of the challenges of deep learning. To ease the preparation of training data, an automated synthesis approach for creating deep learning instance segmentation datasets was introduced in [48]. In this approach, chicken nuggets were used as recognition targets. The nuggets were

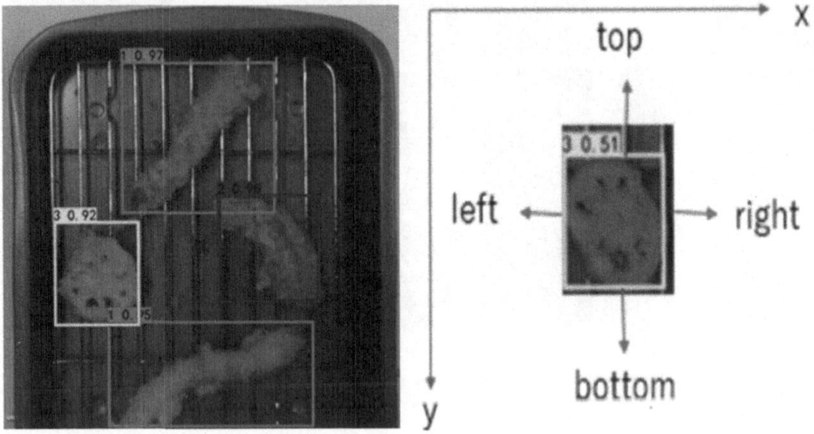

(a) (b)

FIGURE 5.9
Recognition of tempura using YOLO (a) and the resulted bounding box (b).

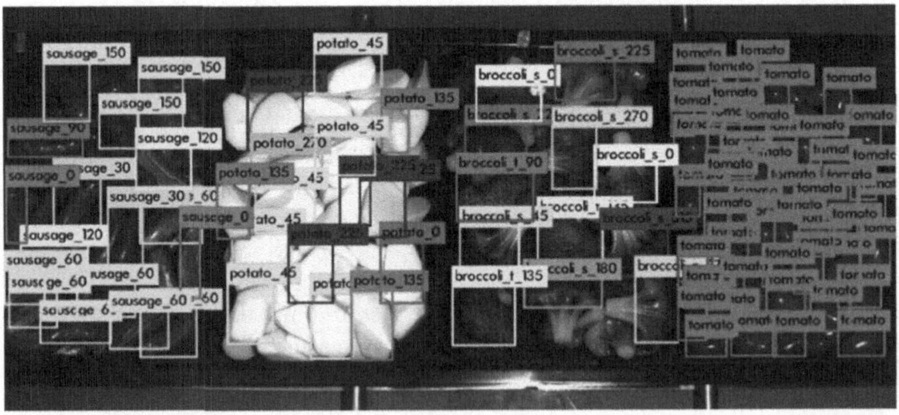

FIGURE 5.10
Recognition example of food bin-picking scenario. Reproduced with permission [47]. Copyright 2022, IEEE.

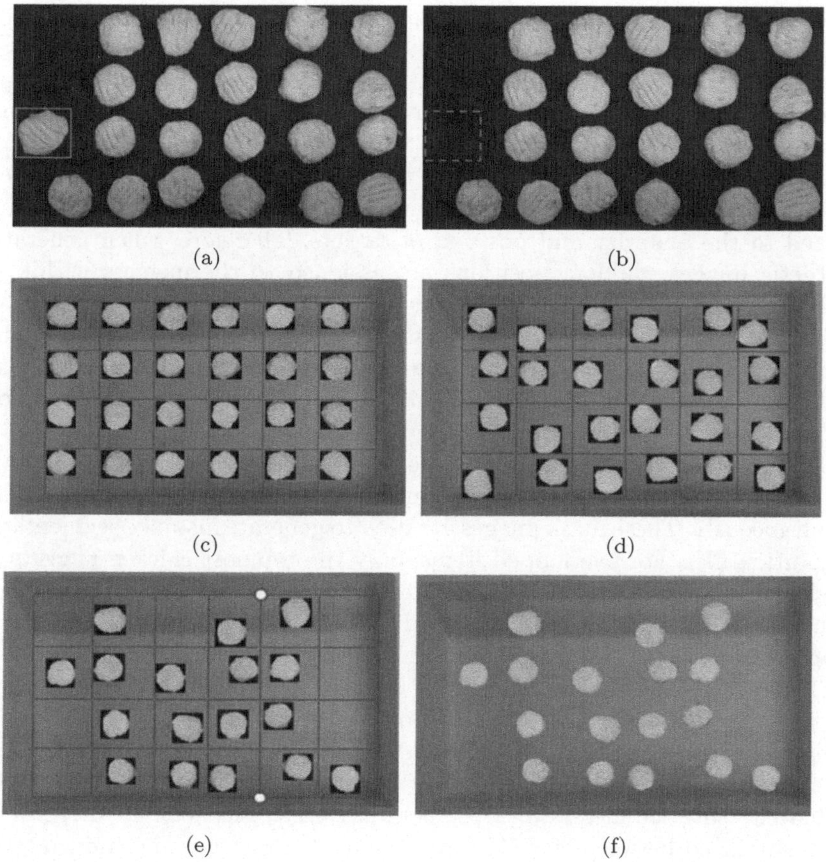

FIGURE 5.11
Automatic data generation for food recognition: (a) the original image, (b) the image after removing one chicken nugget, (c) arranging the chicken nugget images in the container, (d) adding random positions to the nuggets, (e) randomly removing some nuggets, and (f) filling the black pixels with the background color. Reproduced with permission [48]. Copyright 2024, IEEE.

initially placed separately on a background table, as shown in Figure 5.11a, and an image was captured. A chicken nugget (outlined with a dashed rectangle) was then removed from the image, and another image was taken, as shown in Figure 5.11b. By subtracting the second image from the original, a model of the chicken nugget could be obtained. This process was repeated for each nugget, until all chicken nugget models were extracted. These models were then used to generate batches of dataset images and corresponding labels for training.

The synthesis of composite training images is achieved by inserting the previously extracted nugget models into a background image; in this case, a tray. The method begins by generating a grid coordinate system, as shown in Figure 5.11c, and placing the target models on these grids. The nugget model images are arranged uniformly in the top-left corner of each grid. The grids are positioned as close as possible to the lower region of the tray, replicating a real-world scenario. During deep learning, the model can learn features related to the quantity and position of targets. Therefore, when generating synthetic images, random positions are assigned to the nuggets within the grids, ensuring they do not overlap, as shown in Figure 5.11d. To introduce variability in the number of nuggets, some grids are left empty by randomly skipping positions, resulting in the images shown in Figure 5.11e. Finally, the background regions in the images are filled with the tray's background color to reproduce the actual scenario, as seen in Figure 5.11f.

The same method can also be used to generate images with overlapping nuggets. During the dataset generation process, mask images can be created simultaneously. These mask images are used to generate instance segmentation annotation files. For overlapped items, only the topmost chicken nuggets are labeled. Figure 5.12a illustrates an overlapping scenario, where the nuggets highlighted in the circles are not labeled in the mask. The corresponding mask image for the scenario shown in Figure 5.12a is presented in Figure 5.12b.

The YOLACT (You Only Look At CoefficienTs) algorithm was employed for segmentation. Two sets of images, one with and one without overlapping nuggets, were created for training, each consisting of 500 images. The dataset generation process took approximately 3 min. The training performance was evaluated using the mean average precision (mAP) metric. The mAP for the mask stabilized between 90–95% after around 120 epochs. The trained model was then evaluated using the image shown in Figure 5.12c, and the recognition result is shown in Figure 5.12d.

To demonstrate robotic grasping based on nugget recognition, experiments were conducted using a five-degree-of-freedom robotic manipulator (ViperX-300, Robotis) with a needle end-effector to pick up eight overlapped chicken nuggets from a tray and place them onto three desired plates. The experimental setup is shown in Figure 5.13a. The fingers of the needle end-effector are constructed from multiple thin wires (Figure 5.13b), offering a certain level of flexibility that helps minimize potential damage to the nuggets. An RGB-D camera (RealSense D435i, Intel) is mounted at the end of the robotic manipulator to capture images of the nuggets.

The prediction results for picking up chicken nuggets are shown in Figure 5.14. In Figure 5.14a, the arrangement of the chicken nuggets in the food container was prepared in advance. The prediction results of the first handling task are shown in Figure 5.14b. The chicken nugget with the highest confidence score was selected for handling. Once the grabbed target in Figure 5.14b

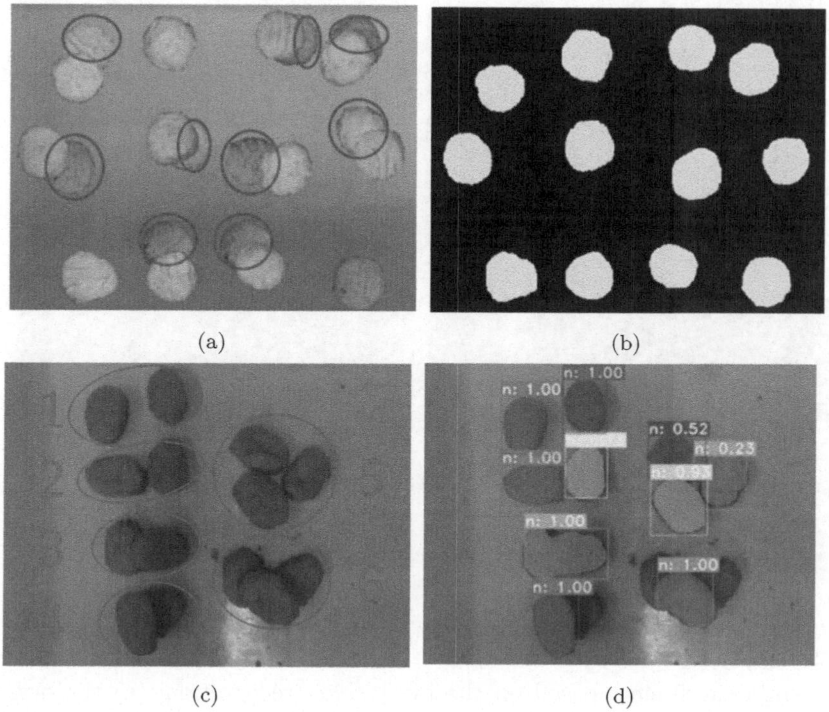

FIGURE 5.12
Data annotation and recognition examples of chicken nuggets: (a) scenario of overlapped chicken nuggets, (b) annotation of chicken nuggets, (c) original image for recognition test, and (d) the recognition results of overlapped chicken nuggets. Reproduced with permission [48]. Copyright 2024, IEEE.

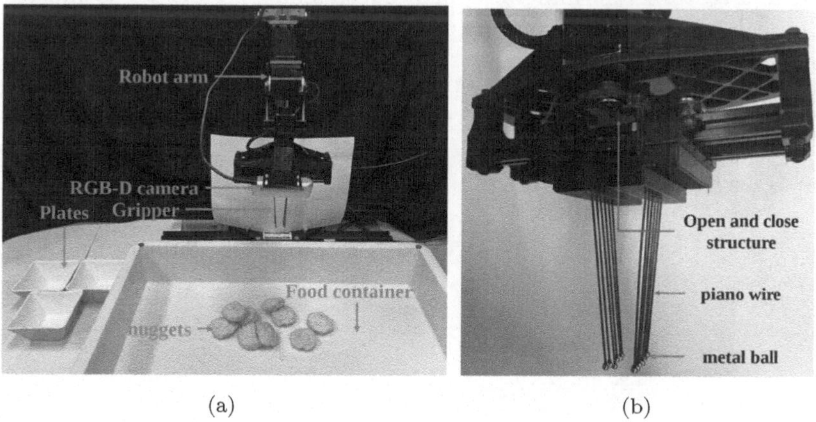

FIGURE 5.13
Experimental setup (a) and the robotic end-effector (b). Reproduced with permission [48]. Copyright 2024, IEEE.

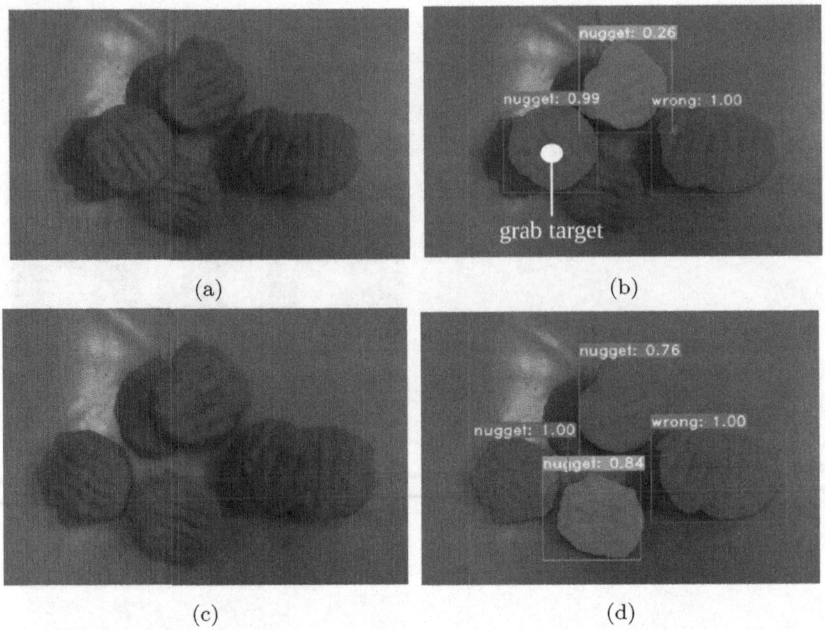

FIGURE 5.14
Grasping experiments based on the bin-picking recognition: (a) the original image, (b) nugget recognition for the first pick, (c) image after the first pick, and (d) nugget recognition for the second pick. Reproduced with permission [48]. Copyright 2024, IEEE.

was removed, the previously obscured nuggets beneath it became visible, as shown in Figure 5.14c, and were then successfully predicted, as illustrated in Figure 5.14d. During experiments, six out of eight chicken nuggets were successfully predicted and handled.

6

Food Modeling and Property Measurements

This chapter focuses on the physical modeling of food items. Based on the measurements of various food products, we develop computational models to simulate their behavior. We discuss how to formulate the deformation of food items and identify model parameters from these measurements. Additionally, we introduce physical twins of food materials, which are made from elastomers and plastics. These physical twins exhibit properties that closely resemble those of real food items. Several examples of food physical twins are presented to demonstrate their application.

6.1 Deformation Modeling

6.1.1 Elastic and viscous models

Food materials exhibit various deformation properties. Let us apply an external force to a food material and release the applied force. Depending on the shape after the release, deformation can be classified into three categories: elastic deformation, plastic deformation, and elastoplastic deformation. Elastic deformation implies that the deformed shape completely returns to the natural shape after the release. Plastic deformation implies that the deformed shape caused by the applied force completely remains even after the release. Elastoplastic deformation implies that the deformation caused by the applied force partially remains.

Deformation properties can be formulated by stress-strain relationships, which are often described by combinations of elastic and viscous models. Elastic models define the relationship between stress and strain. Let σ be the stress applied to an elastic model and ε be the strain caused by the applied stress. Linear elastic model is then formulated as:

$$\sigma = E\varepsilon \tag{6.1}$$

where proportional constant E is referred to as Young's modulus. Linear elastic model is described by a symbol shown in Figure 6.1a. Note that this linear elasticity is ideal; real food materials exhibit nonlinear and time-variant elasticity. Unit of stress σ is N/m^2 or Pa, while strain ε is dimensionless. Thus, unit of Young's modulus E is N/m^2 or Pa.

DOI: 10.1201/9781003372431-6

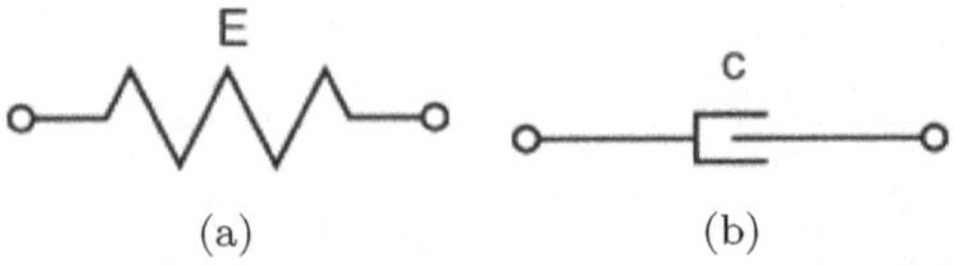

FIGURE 6.1
Fundamental models. (a) Elastic model. (b) Viscous model.

Viscous models define the relationship between stress and the rate of strain. The rate of strain at time t is given by time-derivative of strain; $\dot{\varepsilon} = d\varepsilon/dt$. Linear viscous model is then formulated as:

$$\sigma = c\dot{\varepsilon} \tag{6.2}$$

where proportional constant c is referred to as viscous modulus. Linear viscous model is described by a symbol shown in Figure 6.1b. Note that this linear viscosity is ideal; real food materials exhibit nonlinear and time-variant viscosity. Unit of the rate of strain $\dot{\varepsilon}$ is 1/s, implying that unit of viscous modulus c is $\mathrm{Ns/m^2}$ or $\mathrm{Pa \cdot s}$.

Let us apply a constant stress σ_0 to linear elastic model $\sigma = E\varepsilon$ during time $[\, t_1,\, t_2 \,]$. Strain of the linear elastic model is then described as:

$$\varepsilon = \begin{cases} 0 & t \leq t_1 \\ \sigma_0/E & t \in [\, t_1,\, t_2 \,] \\ 0 & t \geq t_2 \end{cases}$$

This response is shown in Figure 6.2a. Non-zero strain σ_0/E occurs as long as constant stress σ_0 is applied. Strain changes discontinuously at time t_1 and

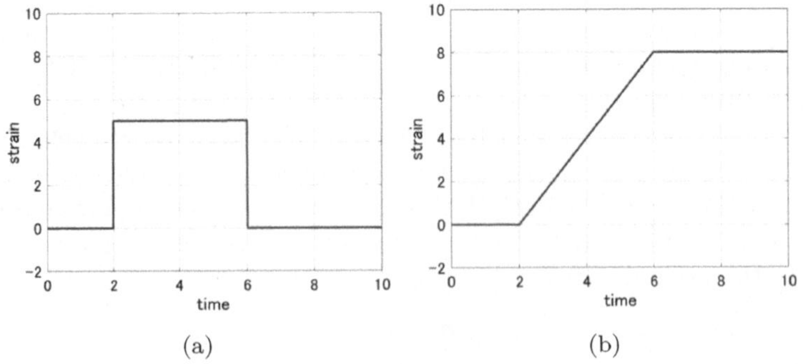

FIGURE 6.2
Response of fundamental models: (a) response of elastic element: $E = 2.0$, $\sigma_0 = 10.0$, $t_1 = 2.0$, $t_2 = 6.0$ and (b) response of viscous element: $c = 5.0$, $\sigma_0 = 10.0$, $t_1 = 2.0$, $t_2 = 6.0$.

t_2. Let us apply a constant stress σ_0 to linear viscous model $\sigma = c\dot{\varepsilon}$ during time $[\,t_1,\, t_2\,]$. Strain of the linear viscous model is then described as:

$$\varepsilon = \begin{cases} 0 & t \le t_1 \\ (\sigma_0/c)(t - t_1) & t \in [\,t_1,\, t_2\,] \\ (\sigma_0/c)(t_2 - t_1) & t \ge t_2 \end{cases}$$

This response is shown in Figure 6.2b. Strain constantly increases at the rate of σ_0/c as long as constant stress σ_0 is applied. After the applied stress is released, constant strain $(\sigma_0/c)(t_2 - t_1)$ remains. Consequently, this viscous model can describe plastic deformation.

6.1.2 Voigt, Maxwell, and three-element models

Voigt model is a parallel connection of a linear elastic element characterized by Young's modulus E and a linear viscous element characterized by viscous modulus c. Voigt model is described by a symbol shown in Figure 6.3a.

Let σ be the stress applied to Voigt model and ε be the strain caused by the applied stress. The total stress σ is the sum of stresses $E\varepsilon$ and $c\dot{\varepsilon}$; the former is caused by the linear elastic element and the latter is caused by the linear viscous element. Thus, we have the following stress-strain relationship:

$$\sigma = E\varepsilon + c\dot{\varepsilon} \tag{6.3}$$

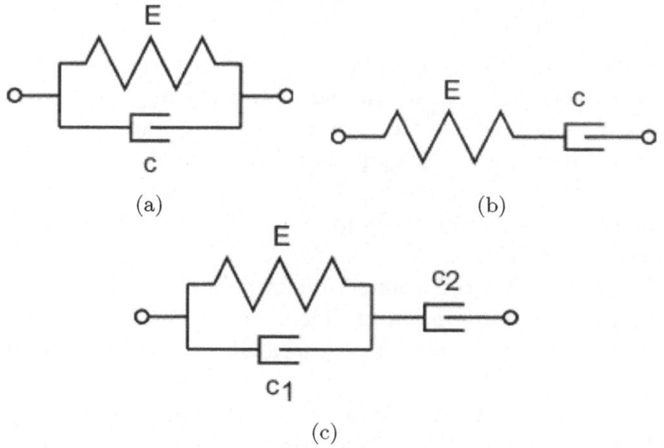

FIGURE 6.3
Combined models. (a) Voigt model. (b) Maxwell model. (c) Three-element model.

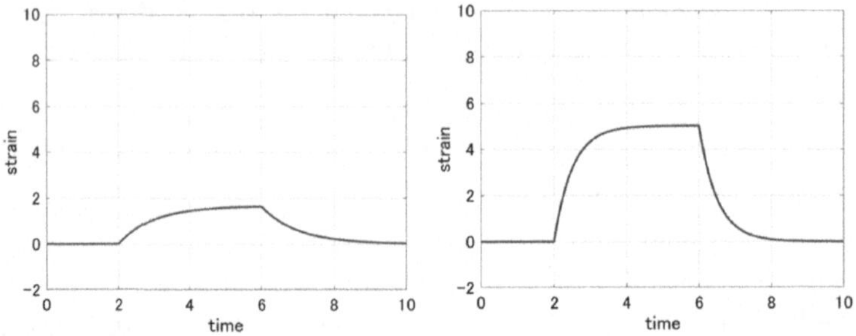

(a) $E = 6.0$, $c = 6.0$ (b) $E = 2.0$, $c = 1.0$

FIGURE 6.4
Response of Voigt model: $\sigma_0 = 10.0$, $t_1 = 2.0$, $t_2 = 6.0$.

This equation is an ordinary differential equation with respect to strain ε of the first order. By solving the above equation, we have

$$\varepsilon(t) = \int_0^t \frac{1}{c} e^{-\frac{E}{c}(t-t')} \sigma(t')\, dt' \tag{6.4}$$

Let us apply constant stress σ_0 to Voigt model during time period $[t_1, t_2]$. Strain of the model is then described as

$$\varepsilon(t) = \begin{cases} 0 & t \leq t_1 \\ \frac{\sigma_0}{E}\{1 - e^{-\frac{E}{c}(t-t_1)}\} & t \in [t_1, t_2] \\ \frac{\sigma_0}{E}\{1 - e^{-\frac{E}{c}(t_2-t_1)}\} e^{-\frac{E}{c}(t-t_2)} & t \geq t_2 \end{cases} \tag{6.5}$$

Strain converges to stationary strain σ_0/E as long as constant stress is applied to the model. Additionally, strain vanishes after the applied stress is released. Ratio E/c determines how fast the strain changes. Figure 6.4 shows the strain response of two Voigt models. Recall that discontinuity in strain for constant stress occurs in linear elastic model. Voigt model solves this problem. Strain of Voigt model changes continuously for constant stress. This Voigt model can describe elastic deformation.

Maxwell model is a serial connection of a linear elastic element characterized by Young's modulus E and a linear viscous element characterized by viscous modulus c. Maxwell model is described by a symbol shown in Figure 6.3b.

Let σ be the stress applied to Maxwell model and ε be the strain caused by the applied stress. Let $\varepsilon_{\mathrm{ela}}$ and $\varepsilon_{\mathrm{vis}}$ be the strains of elastic and viscous elements, yielding $\sigma = E\varepsilon_{\mathrm{ela}}$ and $\sigma = c\dot{\varepsilon}_{\mathrm{vis}}$. Strain ε is the sum of $\varepsilon_{\mathrm{ela}}$ and $\varepsilon_{\mathrm{vis}}$, that is, $\varepsilon = \varepsilon_{\mathrm{ela}} + \varepsilon_{\mathrm{vis}}$. Assuming that $\varepsilon = 0$ and $\sigma = 0$ at $t = 0$, we have the following equation:

$$\varepsilon(t) = \frac{\sigma}{E} + \frac{1}{c}\int_0^t \sigma(t')\, dt' \tag{6.6}$$

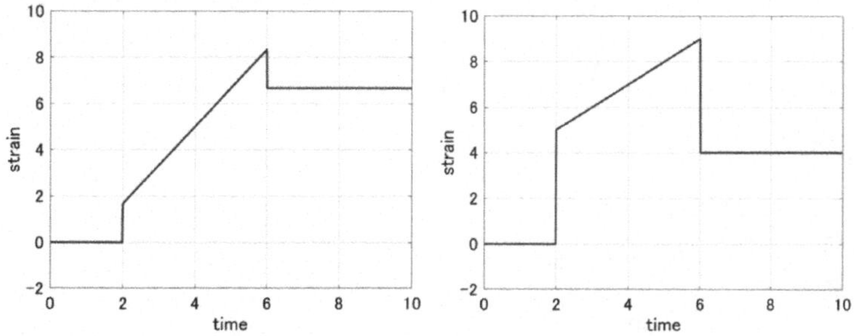

(a) $E = 6.0$, $c = 6.0$ (b) $E = 2.0$, $c = 10.0$

FIGURE 6.5
Response of Maxwell model: $\sigma_0 = 10.0$, $t_1 = 2.0$, $t_2 = 6.0$.

Let us apply constant stress σ_0 to Maxwell model during time period $[t_1, t_2]$. Strain of the model is then described as

$$\varepsilon(t) = \begin{cases} 0 & t \le t_1 \\ \dfrac{\sigma_0}{E} + \dfrac{\sigma_0}{c}(t - t_1) & t \in [t_1, t_2] \\ \dfrac{\sigma_0}{c}(t_2 - t_1) & t \ge t_2 \end{cases} \tag{6.7}$$

Strain increases as long as constant stress is applied. Constant strain $(\sigma_0/c)(t_2 - t_1)$ remains after the applied stress is released. Figure 6.5 shows the strain response of two Maxwell models. Strain changes discontinuously at time t_1 and t_2.

Three-element model consists of a Voigt element and a viscous element connected in serial. Let E and c_1 be Young's modulus and viscous modulus of the Voigt element. Let c_2 be viscous modulus of the other viscous element. Three-element model is described by a symbol shown in Figure 6.3c.

Let σ be the stress applied to three-element model and ε be the strain caused by the applied stress. Let $\varepsilon_{\text{voigt}}$ and ε_{vis} be the strains of Voigt and viscous elements. Then, we have

$$\sigma = E\varepsilon_{\text{voigt}} + c_1\dot{\varepsilon}_{\text{voigt}}$$
$$\sigma = c_2\dot{\varepsilon}_{\text{vis}}$$

Assume that $\varepsilon_{\text{voigt}} = 0$, $\varepsilon_{\text{vis}} = 0$, and $\sigma = 0$ at $t = 0$. Solving the above equations, we find

$$\varepsilon_{\text{voigt}}(t) = \int_0^t \frac{1}{c_1} e^{-\frac{E}{c_1}(t-t')} \sigma(t')\,\mathrm{d}t'$$
$$\varepsilon_{\text{vis}}(t) = \int_0^t \frac{1}{c_2} \sigma(t')\,\mathrm{d}t'$$

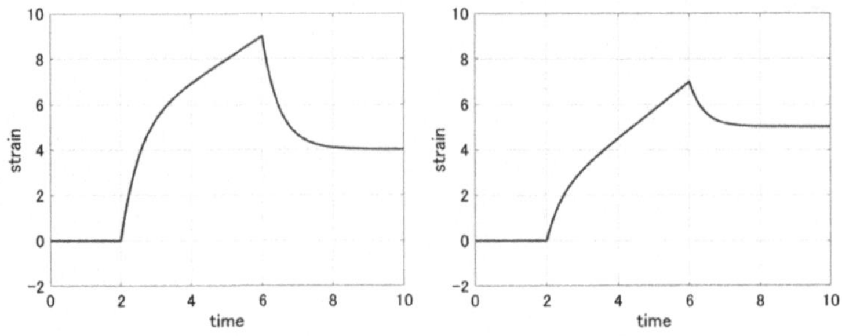

(a) $E = 2.0$, $c_1 = 1.0$, $c_2 = 10.0$ (b) $E = 5.0$, $c_1 = 2.0$, $c_2 = 8.0$

FIGURE 6.6
Response of three-element model: $\sigma_0 = 10.0$, $t_1 = 2.0$, $t_2 = 6.0$.

Strain ε is the sum of $\varepsilon_{\text{voigt}}$ and ε_{vis}. Consequently, we have the following equation:

$$\varepsilon(t) = \int_0^t \frac{1}{c_1} e^{-\frac{E}{c_1}(t-t')} \sigma(t')\,\mathrm{d}t' + \int_0^t \frac{1}{c_2} \sigma(t')\,\mathrm{d}t' \tag{6.8}$$

Let us apply constant stress σ_0 to three-element model during time period $[t_1, t_2]$. Strain of the model is then described as

$$\varepsilon(t) = \begin{cases} 0 & t \le t_1 \\ \dfrac{\sigma_0}{E}\{1 - e^{-\frac{E}{c_1}(t-t_1)}\} + \dfrac{\sigma_0}{c_2}(t - t_1) & t \in [t_1, t_2] \\ \dfrac{\sigma_0}{E}\{1 - e^{-\frac{E}{c_1}(t_2-t_1)}\} e^{-\frac{E}{c}(t-t_2)} + \dfrac{\sigma_0}{c_2}(t_2 - t_1) & t \ge t_2 \end{cases} \tag{6.9}$$

Strain increases as long as constant stress is applied. Strain decreases after the applied stress is released. Note that constant strain $(\sigma_0/c_2)(t_2 - t_1)$ remains when $t \to \infty$. Figure 6.6 shows the strain response of two three-element models. Strain of three-element model changes continuously for constant stress. This three-element model can describe elastoplastic deformation.

6.1.3 Relaxation function

Maxwell model yields the following ordinary differential equation with respect to stress and strain:

$$\dot{\sigma} + \frac{E}{c}\sigma = E\dot{\varepsilon} \tag{6.10}$$

Solving the above equation with respect to ε, we have the following stress-strain relationship in a convolution form:

$$\sigma(t) = \int_0^t r(t - t')\,\dot{\varepsilon}(t')\,\mathrm{d}t' \tag{6.11}$$

where

$$r(t - t') = Ee^{-\frac{E}{c}(t-t')} \tag{6.12}$$

Function $r(t - t')$ is referred to as a relaxation function.

Three-element model yields the following ordinary differential equation with respect to stress and strain:

$$\dot{\sigma} + \frac{E}{c_1 + c_2}\sigma = \frac{Ec_2}{c_1 + c_2}\dot{\varepsilon} + \frac{c_1 c_2}{c_1 + c_2}\ddot{\varepsilon} \tag{6.13}$$

Solution of this equation is also described in equation (6.11) with

$$r(t - t') = \frac{c_2}{c_1 + c_2}e^{-\frac{E}{c_1+c_2}(t-t')}\left(E + c_1\frac{\mathrm{d}}{\mathrm{d}t}\right) \tag{6.14}$$

Note that this relaxation function involves operator $\mathrm{d}/\mathrm{d}t$.

6.1.4 Strain and stress vectors

Stress-strain relationships in Sections 6.1.1 and 6.1.2 are 1D. 1D deformation consists of extensional strain alone. In contrast, 2D or 3D deformations consist of not only extensional deformation but also shear deformation. 2D deformation in xy plane consists of extensional deformations along x- and y-axes as well as shear deformation in xy plane. 3D deformation in xyz space consists of extensional deformations along x-, y-, and z-axes as well as shear deformations in yz, zx, and xy planes. Consequently, 2D deformation is formulated by the following strain components:

$$\begin{array}{rl} \varepsilon_x & \text{extensional strain along } x\text{-axis} \\ \varepsilon_y & \text{extensional strain along } y\text{-axis} \\ \varepsilon_{xy} & \text{shear strain in } xy \text{ plane} \end{array} \tag{6.15}$$

with stress components

$$\begin{array}{rl} \sigma_x & \text{extensional stress along } x\text{-axis} \\ \sigma_y & \text{extensional stress along } y\text{-axis} \\ \sigma_{xy} & \text{shear stress in } xy \text{ plane} \end{array} \tag{6.16}$$

Strain vector $\boldsymbol{\varepsilon}$ and stress vector $\boldsymbol{\sigma}$ are then defined as follows:

$$\boldsymbol{\varepsilon} = \begin{bmatrix} \varepsilon_x \\ \varepsilon_y \\ 2\varepsilon_{xy} \end{bmatrix}, \qquad \boldsymbol{\sigma} = \begin{bmatrix} \sigma_x \\ \sigma_y \\ \sigma_{xy} \end{bmatrix} \tag{6.17}$$

Inner product $\boldsymbol{\sigma} \cdot \boldsymbol{\varepsilon}$ corresponds to work per unit volume. Similarly, 3D deformation is formulated by the following strain components:

$$\begin{array}{rl} \varepsilon_x & \text{extensional strain along } x\text{-axis} \\ \varepsilon_y & \text{extensional strain along } y\text{-axis} \\ \varepsilon_z & \text{extensional strain along } z\text{-axis} \\ \varepsilon_{yz} & \text{shear strain in } yz \text{ plane} \\ \varepsilon_{zx} & \text{shear strain in } zx \text{ plane} \\ \varepsilon_{xy} & \text{shear strain in } xy \text{ plane} \end{array} \tag{6.18}$$

with stress components

$$\begin{array}{rl} \sigma_x & \text{extensional stress along } x\text{-axis} \\ \sigma_y & \text{extensional stress along } y\text{-axis} \\ \sigma_z & \text{extensional stress along } z\text{-axis} \\ \sigma_{yz} & \text{shear stress in } yz \text{ plane} \\ \sigma_{zx} & \text{shear stress in } zx \text{ plane} \\ \sigma_{xy} & \text{shear stress in } xy \text{ plane} \end{array} \tag{6.19}$$

Strain vector $\boldsymbol{\varepsilon}$ and stress vector $\boldsymbol{\sigma}$ are then defined as follows:

$$\boldsymbol{\varepsilon} = \begin{bmatrix} \varepsilon_x \\ \varepsilon_y \\ \varepsilon_z \\ 2\varepsilon_{yz} \\ 2\varepsilon_{zx} \\ 2\varepsilon_{xy} \end{bmatrix}, \quad \boldsymbol{\sigma} = \begin{bmatrix} \sigma_x \\ \sigma_y \\ \sigma_z \\ \sigma_{yz} \\ \sigma_{zx} \\ \sigma_{xy} \end{bmatrix} \tag{6.20}$$

Inner product $\boldsymbol{\sigma} \cdot \boldsymbol{\varepsilon}$ corresponds to work per unit volume.

6.1.5 Multi-dimensional models

Let us formulate stress-strain relationship of a 2D elastic body. Assume that the body material shows linear isotropic elasticity. Isotropy implies that a material shows the same deformation property at any direction. The strain-strain relationship is then formulated as follows:

$$\boldsymbol{\sigma} = (\lambda I_\lambda + \mu I_\mu)\boldsymbol{\varepsilon} \tag{6.21}$$

where λ and μ are Lamé's constants and

$$I_\lambda = \begin{bmatrix} 1 & 1 & \\ 1 & 1 & \\ & & \end{bmatrix}, \quad I_\mu = \begin{bmatrix} 2 & & \\ & 2 & \\ & & 1 \end{bmatrix} \tag{6.22}$$

Lamé's constants λ and μ are specific to body material, and are related to Young's modulus E and Poisson's ratio ν by

$$\lambda = \frac{\nu E}{(1+\nu)(1-2\nu)}, \qquad \mu = \frac{E}{2(1+\nu)} \tag{6.23}$$

Note that tensile test provides the values of Young's modulus E and Poisson's ratio ν, directly yielding the values of Lamé's constants λ and μ by the above equations. Stress-strain relationship of a 3D elastic body is also formulated in the above except

$$I_\lambda = \left[\begin{array}{ccc|ccc} 1 & 1 & 1 & & & \\ 1 & 1 & 1 & & & \\ 1 & 1 & 1 & & & \\ \hline & & & & & \\ & & & & & \\ & & & & & \end{array}\right], \quad I_\mu = \left[\begin{array}{ccc|ccc} 2 & & & & & \\ & 2 & & & & \\ & & 2 & & & \\ \hline & & & 1 & & \\ & & & & 1 & \\ & & & & & 1 \end{array}\right] \tag{6.24}$$

Matrices I_λ and I_μ originate from isotropy.

Recall that the stress-strain relationship of Maxwell or three-element model can be described by a convolution form equation (6.11) with a relaxation function $r(t'-t)$. Then, 2D or 3D elastoplastic deformation can be described as follows:

$$\boldsymbol{\sigma}(t) = \int_0^t R(t-t')\, \dot{\boldsymbol{\varepsilon}}(t')\, \mathrm{d}t', \tag{6.25}$$

where 3×3 or 6×6 matrix $R(t-t')$ determines the nature of a 2D or 3D elastoplastic deformation. Replacing two elastic constants in equation (6.21) by two relaxation functions yields matrix $R(t-t')$ of 2D or 3D isotropic elastoplastic deformation:

$$R(t-t') = r_\lambda(t-t')I_\lambda + r_\mu(t-t')I_\mu \tag{6.26}$$

Recalling relaxation function equation (6.12) in 1D Maxwell model, two relaxation functions $r_\lambda(t-t')$ and $r_\mu(t-t')$ in 2D or 3D isotropic Maxwell model turn into:

$$r_\lambda(t-t') = \lambda^{\mathrm{ela}} \exp\left\{-\frac{\lambda^{\mathrm{ela}}}{\lambda^{\mathrm{vis}}}(t-t')\right\} \tag{6.27a}$$

$$r_\mu(t-t') = \mu^{\mathrm{ela}} \exp\left\{-\frac{\mu^{\mathrm{ela}}}{\mu^{\mathrm{vis}}}(t-t')\right\} \tag{6.27b}$$

Constants $\lambda^{\rm ela}$ and $\mu^{\rm ela}$ specify elasticity of the material while $\lambda^{\rm vis}$ and $\mu^{\rm vis}$ describe its plasticity. Recalling relaxation function given in equation (6.14) in 1D three-element model, two relaxation functions $r_\lambda(t-t')$ and $r_\mu(t-t')$ in 2D or 3D isotropic three-element model turn into:

$$r_\lambda(t-t') = \frac{\lambda_2^{\rm vis}}{\lambda_1^{\rm vis}+\lambda_2^{\rm vis}} \exp\left\{-\frac{\lambda^{\rm ela}}{\lambda_1^{\rm vis}+\lambda_2^{\rm vis}}(t-t')\right\}\left(\lambda^{\rm ela}+\lambda_1^{\rm vis}\frac{\rm d}{{\rm d}t}\right) \tag{6.28a}$$

$$r_\mu(t-t') = \frac{\mu_2^{\rm vis}}{\mu_1^{\rm vis}+\mu_2^{\rm vis}} \exp\left\{-\frac{\mu^{\rm ela}}{\mu_1^{\rm vis}+\mu_2^{\rm vis}}(t-t')\right\}\left(\mu^{\rm ela}+\mu_1^{\rm vis}\frac{\rm d}{{\rm d}t}\right) \tag{6.28b}$$

Constants $\lambda^{\rm ela}$ and $\mu^{\rm ela}$ specify elasticity of the material, $\lambda_1^{\rm vis}$ and $\mu_1^{\rm vis}$ describe its viscosity, and $\lambda_2^{\rm vis}$ and $\mu_2^{\rm vis}$ show its plasticity.

6.1.6 Finite element formulation

Finite element formulation provides a numerical method to compute deformation of soft bodies. In finite element formulation, region of a soft body is divided into a finite number of small regions. A finite number of points, which are referred to as nodal points, are selected to describe the deformation of body. Stress-strain relationships are converted into relations between forces at the nodal points and displacements of the nodal points, which contribute to the computation of deformation.

Figure 6.7 shows two examples of finite element approximation of a rectangle of width 2 and height 1. The rectangle is divided into four triangles in Figure 6.7a while two squares in Figure 6.7b. Six nodal points $\rm P_1$–$\rm P_6$ are selected in both. Let a 2D vector $\boldsymbol{u}_k$ be displacement of point $\rm P_k$ and $\boldsymbol{f}_k$ be a 2D vector of force at the points. Let $\boldsymbol{u}_{\rm N}$ be a collective vector of $\boldsymbol{u}_1$–$\boldsymbol{u}_6$, which is referred to as nodal displacement vector. Let $\boldsymbol{f}_{\rm N}$ be a collective vector of

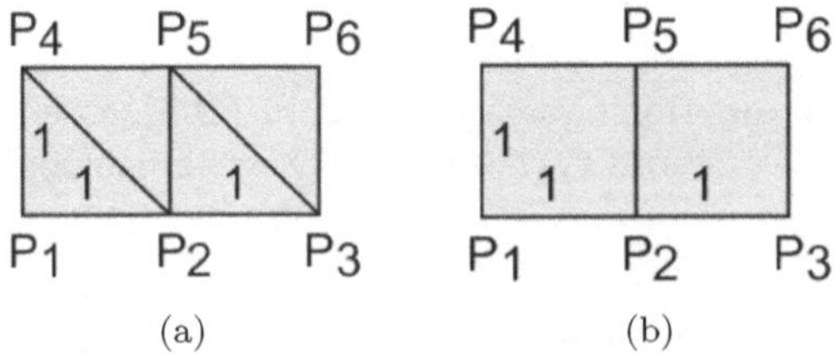

FIGURE 6.7
Examples of finite element approximation. (a) Triangles. (b) Squares.

$\boldsymbol{f}_1$–$\boldsymbol{f}_6$, which is referred to as nodal force vector. Stress-strain relationship of elastic material given in Figure 6.1a is then converted into the relationship between nodal force vector and nodal displacement vector:

$$\boldsymbol{f}_{\mathrm{N}} = -(\lambda J_\lambda + \mu J_\mu)\, \boldsymbol{u}_{\mathrm{N}} \tag{6.29}$$

where J_λ and J_μ are 12×12 matrices depending on finite element division and nodal point coordinates. In Figure 6.7a, these matrices are

$$J_\lambda = \frac{1}{2}\left[\begin{array}{cc|cc|cc|cc|cc|cc}
1 & 1 & -1 & 0 & & & 0 & -1 & & & & \\
1 & 1 & -1 & 0 & & & 0 & -1 & & & & \\ \hline
-1 & -1 & 2 & 1 & -1 & 0 & 0 & 1 & 0 & -1 & & \\
0 & 0 & 1 & 2 & -1 & 0 & 1 & 0 & -1 & -2 & & \\ \hline
 & & -1 & -1 & 1 & 0 & & & 0 & 1 & 0 & 0 \\
 & & 0 & 0 & 0 & 1 & & & 1 & 0 & -1 & -1 \\ \hline
0 & 0 & 0 & 1 & & & 1 & 0 & -1 & -1 & & \\
-1 & -1 & 1 & 0 & & & 0 & 1 & 0 & 0 & & \\ \hline
 & & 0 & -1 & 0 & 1 & -1 & 0 & 2 & 1 & -1 & -1 \\
 & & -1 & -2 & 1 & 0 & -1 & 0 & 1 & 2 & 0 & 0 \\ \hline
 & & & & 0 & -1 & & & -1 & 0 & 1 & 1 \\
 & & & & 0 & -1 & & & -1 & 0 & 1 & 1
\end{array}\right]$$

$$J_\mu = \frac{1}{2}\left[\begin{array}{cc|cc|cc|cc|cc|cc}
3 & 1 & -2 & -1 & & & -1 & 0 & & & & \\
1 & 3 & 0 & -1 & & & -1 & -2 & & & & \\ \hline
-2 & 0 & 6 & 1 & -2 & -1 & 0 & 1 & -2 & -1 & & \\
-1 & -1 & 1 & 6 & 0 & -1 & 1 & 0 & -1 & -4 & & \\ \hline
 & & -2 & 0 & 3 & 0 & & & 0 & 1 & -1 & -1 \\
 & & -1 & -1 & 0 & 3 & & & 1 & 0 & 0 & -2 \\ \hline
-1 & -1 & 0 & 1 & & & 3 & 0 & -2 & 0 & & \\
0 & -2 & 1 & 0 & & & 0 & 3 & -1 & -1 & & \\ \hline
 & & -2 & -1 & 0 & 1 & -2 & -1 & 6 & 1 & -2 & 0 \\
 & & -1 & -4 & 1 & 0 & 0 & -1 & 1 & 6 & -1 & -1 \\ \hline
 & & & & -1 & 0 & & & -2 & -1 & 3 & 1 \\
 & & & & -1 & -2 & & & 0 & -1 & 1 & 3
\end{array}\right]$$

In Figure 6.7b, these matrices are

$$
J_\lambda = \frac{1}{12}\left[\begin{array}{cc|cc|cc|cc|cc|cc}
4 & 3 & -4 & 3 & & & 2 & -3 & -2 & -3 & & \\
3 & 4 & -3 & 2 & & & 3 & -4 & -3 & -2 & & \\
\hline
-4 & -3 & 8 & 0 & -4 & 3 & -2 & 3 & 4 & 0 & -2 & -3 \\
3 & 2 & 0 & 8 & -3 & 2 & 3 & -2 & 0 & -8 & -3 & -2 \\
\hline
 & & -4 & -3 & 4 & -3 & & & -2 & 3 & 2 & 3 \\
 & & 3 & 2 & -3 & 4 & & & 3 & -2 & -3 & -4 \\
\hline
2 & 3 & -2 & 3 & & & 4 & -3 & -4 & -3 & & \\
-3 & -4 & 3 & -2 & & & -3 & 4 & 3 & 2 & & \\
\hline
-2 & -3 & 4 & 0 & -2 & 3 & -4 & 3 & 8 & 0 & -4 & -3 \\
-3 & -2 & 0 & -8 & 3 & -2 & -3 & 2 & 0 & 8 & 3 & 2 \\
\hline
 & & -2 & -3 & 2 & -3 & & & -4 & 3 & 4 & 3 \\
 & & -3 & -2 & 3 & -4 & & & -3 & 2 & 3 & 4
\end{array}\right]
$$

$$
J_\mu = \frac{1}{4}\left[\begin{array}{cc|cc|cc|cc|cc|cc}
4 & 1 & -2 & -1 & & & 0 & 1 & -2 & -1 & & \\
1 & 4 & 1 & 0 & & & -1 & -2 & -1 & -2 & & \\
\hline
-2 & 1 & 8 & 0 & -2 & -1 & -2 & 1 & 0 & 0 & -2 & -1 \\
-1 & 0 & 0 & 8 & 1 & 0 & 1 & -2 & 0 & -4 & -1 & -2 \\
\hline
 & & -2 & 1 & 4 & -1 & & & -2 & 1 & 0 & -1 \\
 & & -1 & 0 & -1 & 4 & & & 1 & -2 & 1 & -2 \\
\hline
0 & -1 & -2 & 1 & & & 4 & -1 & -2 & 1 & & \\
1 & -2 & 1 & -2 & & & -1 & 4 & -1 & 0 & & \\
\hline
-2 & -1 & 0 & 0 & -2 & 1 & -2 & -1 & 8 & 0 & -2 & 1 \\
-1 & -2 & 0 & -4 & 1 & -2 & 1 & 0 & 0 & 8 & -1 & 0 \\
\hline
 & & -2 & -1 & 0 & 1 & & & -2 & -1 & 4 & 1 \\
 & & -1 & -2 & -1 & -2 & & & 1 & 0 & 1 & 4
\end{array}\right]
$$

Consequently, replacing $\boldsymbol{\varepsilon}$ by $\boldsymbol{u}_{\mathrm{N}}$, $\boldsymbol{\sigma}$ by $\boldsymbol{f}_{\mathrm{N}}$, I_λ by J_λ, and I_μ by J_μ in the stress-strain relationship, equation (6.21) of elastic material yields the relationship between the nodal elastic force vector and the nodal displacement vector.

Applying the same replacement to 2D or 3D stress-strain relationship in isotropic elastoplastic deformation, which is formulated by equations (6.25) and (6.26), we find

$$\boldsymbol{f}_{\mathrm{N}} = -J_\lambda \boldsymbol{f}_\lambda - J_\mu \boldsymbol{f}_\mu \tag{6.30}$$

and

$$\boldsymbol{f}_\lambda(t) = \int_0^t r_\lambda(t-t')\,\dot{\boldsymbol{u}}_{\mathrm{N}}(t')\,\mathrm{d}t' \tag{6.31a}$$

$$\boldsymbol{f}_\mu(t) = \int_0^t r_\mu(t-t')\,\dot{\boldsymbol{u}}_{\mathrm{N}}(t')\,\mathrm{d}t' \tag{6.31b}$$

Applying equations (6.27a) and (6.27b) to equations (6.31a) and (6.31b) and differentiating the obtained equations, we have

$$\dot{\boldsymbol{f}}_\lambda = -\frac{\lambda^{\rm ela}}{\lambda^{\rm vis}}\boldsymbol{f}_\lambda + \lambda^{\rm ela}\dot{\boldsymbol{u}}_{\rm N} \tag{6.32a}$$

$$\dot{\boldsymbol{f}}_\mu = -\frac{\mu^{\rm ela}}{\mu^{\rm vis}}\boldsymbol{f}_\mu + \mu^{\rm ela}\dot{\boldsymbol{u}}_{\rm N} \tag{6.32b}$$

Note that the above two ordinary differential equations have the same form as equation (6.10). Replacing σ, ε, E, and c in equation (6.10) by $\boldsymbol{f}_\lambda$, $\boldsymbol{u}_{\rm N}$, $\lambda^{\rm ela}$, and $\lambda^{\rm vis}$ yields equation (6.32a). Additionally, replacing σ, ε, E, and c in equation (6.10) by $\boldsymbol{f}_\mu$, $\boldsymbol{u}_{\rm N}$, $\mu^{\rm ela}$, and $\mu^{\rm vis}$ yields equation (6.32b). Consequently, stress-strain relationship yields two ordinary differential equations with respect to $\boldsymbol{f}_\lambda$ and $\boldsymbol{f}_\mu$. Then, nodal force vector is given by equation (6.30).

Similarly, applying equations (6.28a) and (6.28b) to equations (6.31a) and (6.31b) and differentiating the obtained equations, we have

$$\dot{\boldsymbol{f}}_\lambda = -\frac{\lambda^{\rm ela}}{\lambda_1^{\rm vis}+\lambda_2^{\rm vis}}\boldsymbol{f}_\lambda + \frac{\lambda^{\rm ela}\lambda_2^{\rm vis}}{\lambda_1^{\rm vis}+\lambda_2^{\rm vis}}\dot{\boldsymbol{u}}_{\rm N} + \frac{\lambda_1^{\rm vis}\lambda_2^{\rm vis}}{\lambda_1^{\rm vis}+\lambda_2^{\rm vis}}\ddot{\boldsymbol{u}}_{\rm N} \tag{6.33a}$$

$$\dot{\boldsymbol{f}}_\mu = -\frac{\mu^{\rm ela}}{\mu_1^{\rm vis}+\mu_2^{\rm vis}}\boldsymbol{f}_\mu + \frac{\mu^{\rm ela}\mu_2^{\rm vis}}{\mu_1^{\rm vis}+\mu_2^{\rm vis}}\dot{\boldsymbol{u}}_{\rm N} + \frac{\mu_1^{\rm vis}\mu_2^{\rm vis}}{\mu_1^{\rm vis}+\mu_2^{\rm vis}}\ddot{\boldsymbol{u}}_{\rm N} \tag{6.33b}$$

Note that the above two ordinary differential equations have the same form as equation (6.13). Replacing σ, ε, E, c_1, and c_2 in equation (6.13) by $\boldsymbol{f}_\lambda$, $\boldsymbol{u}_{\rm N}$, $\lambda^{\rm ela}$, $\lambda_1^{\rm vis}$, and $\lambda_2^{\rm vis}$ yields equation (6.33a). Additionally, replacing σ, ε, E, c_1, and c_2 in equation (6.13) by $\boldsymbol{f}_\mu$, $\boldsymbol{u}_{\rm N}$, $\mu^{\rm ela}$, $\mu_1^{\rm vis}$, and $\mu_2^{\rm vis}$ yields equation (6.33b). Consequently, stress-strain relationship yields two ordinary differential equations with respect to $\boldsymbol{f}_\lambda$ and $\boldsymbol{f}_\mu$. Then, nodal force vector is given by equation (6.30).

Figure 6.8 shows calculation of deformation of an elastoplastic square body of width 30 cm. The body was divided into $9 \times 9 \times 2$ triangles. Deformation was then described by a set of displacement vectors at 10×10 nodal points. Material property was described by a three-element model with $E = 1.0\,\mathrm{MPa}$, $c_1 = 40\,\mathrm{Pa \cdot s}$, $c_2 = 2.0\,\mathrm{MPa \cdot s}$, $\nu = 0.48$, and density $\rho = 1.0\,\mathrm{g/cm^3}$. During time interval $[\,0,\,t_{\rm push}\,]$, the bottom of the body is fixed to the floor and the top of the body $\mathrm{P}_{14}\mathrm{P}_{15}$ moves downward at a constant velocity $v_{\rm push}$. During $[\,t_{\rm push},\,t_{\rm hold}\,]$, the bottom remains fixed and and the top keeps its position. During $[\,t_{\rm hold},\,t_{\rm end}\,]$, the bottom remains fixed while the top is released. We used $t_{\rm push} = 1.0\,\mathrm{s}$, $t_{\rm hold} = 2.0\,\mathrm{s}$, and $v_{\rm push} = 16\,\mathrm{cm/s}$. Deformation remains even after pushing motion is released.

Figure 6.9 shows calculation of deformation of a layered body consisting of two materials. Dark region corresponds to a harder material specified by $E = 1.0\,\mathrm{MPa}$, $c_1 = 40\,\mathrm{Pa \cdot s}$, $c_2 = 2.0\,\mathrm{MPa \cdot s}$, $\nu = 0.48$, and $\rho = 1.0\,\mathrm{g/cm^3}$, while light region corresponds to a softer material specified by $E = 0.2\,\mathrm{MPa}$, $c_1 = 40\,\mathrm{Pa \cdot s}$, $c_2 = 2.0\,\mathrm{MPa \cdot s}$, $\nu = 0.48$, and $\rho = 1.0\mathrm{g/cm^3}$. Thus, material in dark region is five times harder than the material in light region. We find

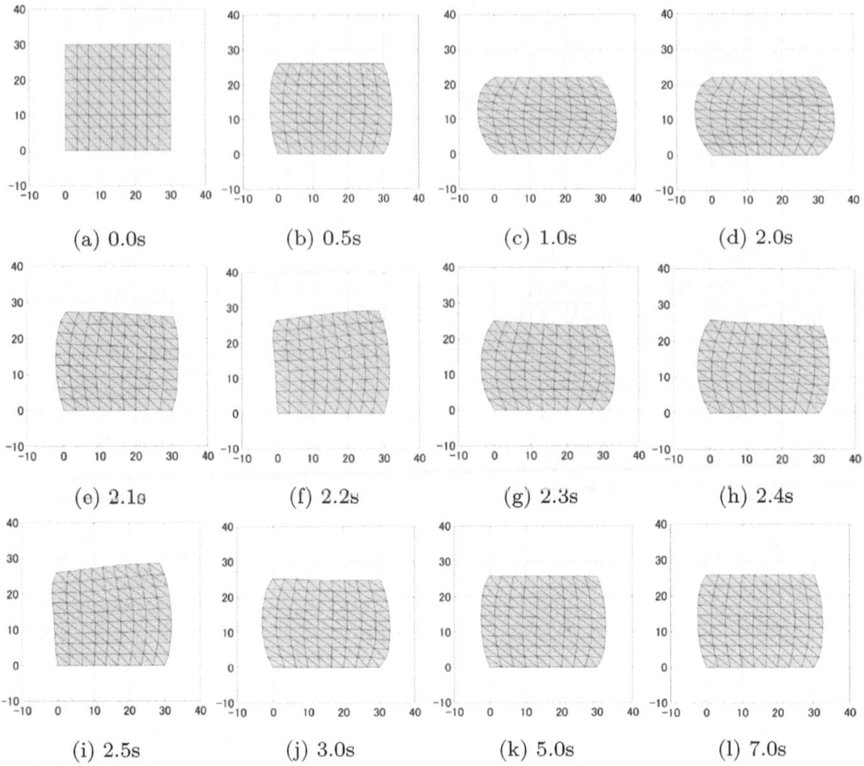

FIGURE 6.8
Computed deformation of elastoplastic square body.

that the softer layer (light region) deforms more than the harder layers (dark region). This deformation of a layered body is different from deformation of a uniform body (Figure 6.8). The softer layer expands outward more in this deformation.

6.2 Property Measurements

6.2.1 Elasticity and viscosity

Elasticity and viscosity are key factors that influence the grasping forces generated when robotic end-effectors interact with food items. By understanding the elasticity and viscosity of food, it becomes possible to estimate the grasping force using an appropriate mechanical model. In the following subsections, we present various models of elasticity and viscosity, along with different

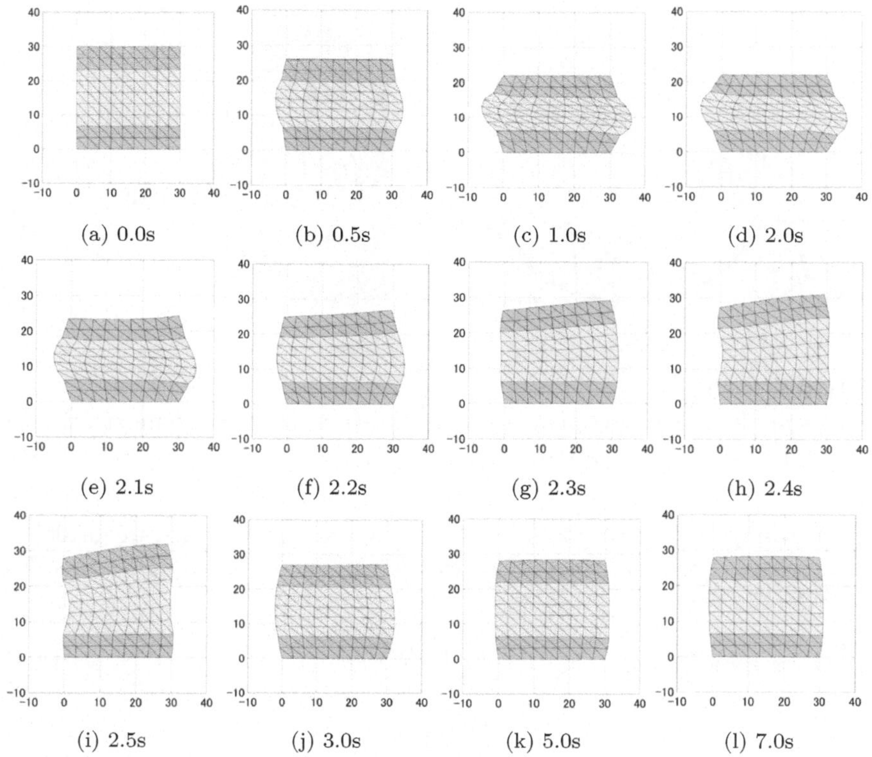

FIGURE 6.9
Computed deformation of layered elastoplastic square body.

approaches for measuring or identifying the physical parameters associated with these models.

6.2.1.1 Measuring Young's modulus

When grasping food items at relatively low speeds, the effect of viscosity can be neglected, and the grasping force primarily depends on the elasticity of food. To characterize the elasticity or hardness of soft materials, Young's modulus is commonly used. To measure Young's modulus, standard tests such as compression or tensile tests are typically performed with standard samples, like cylinders for compression tests and dumbbell specimens for tensile tests. However, food items often have complex geometries, making it difficult to prepare standard test samples. Although test samples can be prepared by cutting a food item into a standard shape, this process may significantly alter the physical properties of the food. To address this issue, we propose modeling the contact between a sphere and the relatively flat surface of a food item using Hertz's contact mechanics [49].

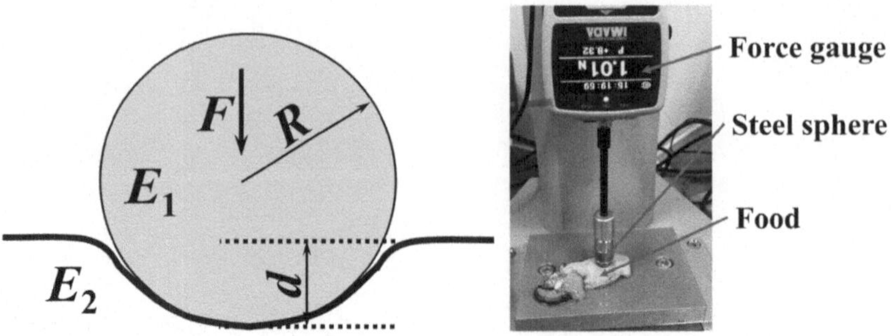

(a) (b)

FIGURE 6.10
Young's modulus measurement using indentation: (a) measurement principle and (b) measurement scenario.

According to Hertz's contact mechanics, as shown in Figure 6.10a, an elastic sphere of radius R indents an elastic half-space. The applied force F is related to the displacement d by

$$F = \frac{4}{3}E^*R^{\frac{1}{2}}d^{\frac{3}{2}} \tag{6.34}$$

where

$$\frac{1}{E^*} = \frac{1-\nu_1^2}{E_1} + \frac{1-\nu_2^2}{E_2}$$

and E_1, E_2 are the Young's moduli and ν_1, ν_2 are the Poisson's ratios associated with sphere and target object, respectively. Considering the scenario of using a metal sphere to indent a soft food item, we have E_1 that is much larger than E_2. Therefore, we have

$$E^* = \frac{E_2}{1-\nu_2^2}.$$

As a result, equation (6.34) turns into the following equation for describing metal sphere indenting into soft food items [50].

$$F = \frac{4}{3}\frac{E}{1-\nu^2}R^{\frac{1}{2}}d^{\frac{3}{2}} \tag{6.35}$$

By knowing the indentation displacement d, Poisson's ratio ν, and the reaction force F, the Young's modulus E can be obtained using curve fitting with the above equation. Note that this equation is only applicable when the indentation is conducted on a relatively flat surface of food. To better satisfy this condition, a smaller metal sphere is usually adopted. Figure 6.10b demonstrates an example of using a test machine (FSA-1KE-50N, IMADA) to measure the Young's modulus of a raw oyster. As examples, we have measured the Young's moduli of food items shown in Figure 6.11 and the measured values are listed in Table 6.1.

FIGURE 6.11
Food items for Young's modulus measurements.

6.2.1.2 Measuring viscoelasticity

When considering high-speed grasping, viscosity of food items significantly affects the grasping force. For food items that recover their deformation after grasping, viscoelastic models are usually used to describe the relationship between force and displacement. For better reproducing the complex force behavior, a model with three spring and damper elements can be applied, as shown in Figure 6.12a. This model has the following governing equation [51]:

$$\dot{f} + \frac{K_2}{B} f = (K_1 + K_2)\dot{x} + \frac{K_1 K_2}{B} x, \tag{6.36}$$

where x and f indicate the displacement and force, K_1 and K_2 the elasticities of the two elastic elements, and B the viscosity of the food item. A typical

TABLE 6.1
Young's modulus of various food items

Food item	Young's modulus [MPa]	Food item	Young's modulus [MPa]
Daifuku	0.01	Boiled egg	0.05
Horse mackerel	0.14	6P cheese	0.08
Conger eel sushi	0.01	Fried shrimp	0.38
Kamaboko	0.05	Fried chicken	0.02
Cucumber	0.99	Mini tomato	0.41
Green pepper	0.17	Omelet	0.03

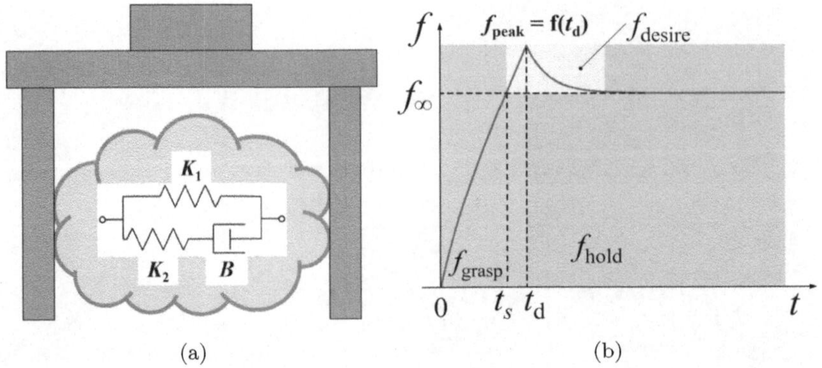

FIGURE 6.12
Model of viscoelasticity (a) and the typical profile of viscoelastic force (b). Reproduced with permission [52]. Copyright 2021, Elsevier.

profile of the viscoelastic force generated by the food item when applying a grasping and holding motion to the food is plotted in Figure 6.12b. While grasping, the force f_{grasp} (region before time t_d in Figure 6.12b) quickly increases and reaches its peak value (f_{peak}) at time t_d. Once the grasping was ended, the deformation was maintained and the force f_{hold} (region after time t_d in Figure 6.12b) started to decrease and finally rested at a constant force f_∞. This force reduction behavior reflects the stress relaxation in viscoelastic objects. The force region above f_∞ (f_{desire}) in Figure 6.12b has force values larger than the steady force (f_∞), but the displacement is equal or less than d. At time t_s, the robotic hand is closing but the force reaches f_∞. Therefore, we can start the next handling motion at time t_s and do not have to wait until the hand completely closed at time t_d. By doing so, we can shorten the time of a handling task. Short handling time is usually preferable for food manufacturing automation. To determine this f_{desire} region, we need to know the food item's viscoelasticity. If we assume that the grasping was conducted with a constant grasping velocity of $\dot{x} = d/t_d$ during the time period $0 < t \leq t_d$, equation (6.36) turns into

$$\dot{f}_{grasp} + \frac{K_2}{B} f_{grasp} = (K_1 + K_2)\frac{d}{t_d} + \frac{K_1 K_2}{B}\frac{d}{t_d}t, \tag{6.37}$$

where d is the predetermined displacement or deformation for grasping. Solving equation (6.37) with initial condition of $f_{grasp} = 0$ at $t = 0$, we can obtain

$$f_{grasp} = (B + K_1 t - Be^{-\frac{K_2}{B}t})\frac{d}{t_d}. \tag{6.38}$$

Similarly, during the time period $t > t_d$, the deformation remains constant with $x = d$. Therefore, equation (6.36) turns into

$$\dot{f}_{hold} + \frac{K_2}{B} f_{hold} = \frac{K_1 K_2}{B} d. \tag{6.39}$$

By solving equation (6.39) with the initial condition of $f_hold = f(t_d)$ at $t = t_d$, we obtain

$$f_{hold} = K_1 d - [K_1 d - f(t_d)]e^{-\frac{K_2}{B}(t-t_d)}. \qquad (6.40)$$

Finally, by knowing the parameters K_1, K_2, and B, the force response from a viscoelastic food item can be calculated using equations (6.38) and (6.40).

To estimate the unknown parameters K_1, K_2, and B, a three-step approach was proposed in [52]. It is assumed that the force data shown in Figure 6.12b are available together with the deformation d of the grasping operation. These data can be easily obtained through a grasp-and-hold experiment using a predetermined either force or deformation as long as the velocity during grasping is kept constant. Owing to the serial connection between the spring element K_2 and the damper element B, the deformation generated in K_2 is gradually absorbed by damper B, due to which, it will eventually vanish at the end of the holding operation. Therefore, the residual force f_∞ is only dependent on the spring element K_1, and we can easily calculate parameter K_1 as

$$K_1 = \frac{f_\infty}{d}.$$

This completes the first step of the parameter estimation. In the second step, we fit the force relaxation data obtained during the holding operation (f_{hold} in Figure 6.12b) using equation (6.40), in which the ratio of $\frac{K_2}{B}$ is the only unknown part. Therefore, this ratio can be determined during the second step. Finally, in the third step, the peak force $f(t_d)$ calculated using equation (6.40) is substituted into equation (6.38), in which, after the above two steps, the only unknown parameter is B. Therefore, parameter B can be calculated and finally parameter K_2 can be determined using the estimated ratio $\frac{K_2}{B}$. Using this three-step approach, one can easily estimate the three viscoelastic parameters of K_2, K_2, and B based on the force measurement of a one-time grasp-and-hold operation.

To efficiently measure the viscoelasticity and friction coefficient of food items, a measuring device was proposed in [52], as shown in Figure 6.13. The device consists of a robotic hand for grasping, two linear stages for vertical translation of the robotic hand and a block bar, a triangle-shaped block bar, and aluminum frames (Figure 6.13a). The robotic hand is constructed using a DC motor (DCX26LGBKL24V, Maxon, Switzerland), a rack and pinion mechanism, two fingers, and a six-axis force sensor (CFS018CA101U, Leptrino, Japan) placed under the fingertip (Figure 6.13b). Additionally, the fingertip includes a slide slot, enabling easy replacement of the fingertip surface. The robotic hand supports both point-to-point (PTP) control and continuous path (CP) control, utilizing feedback from the motor encoder (ENX10EASY1024IMP, Maxon, Switzerland). PTP control is used to adjust the initial grasping position of the fingertip, while CP control is employed to generate the desired indentation path for measuring viscoelasticity.

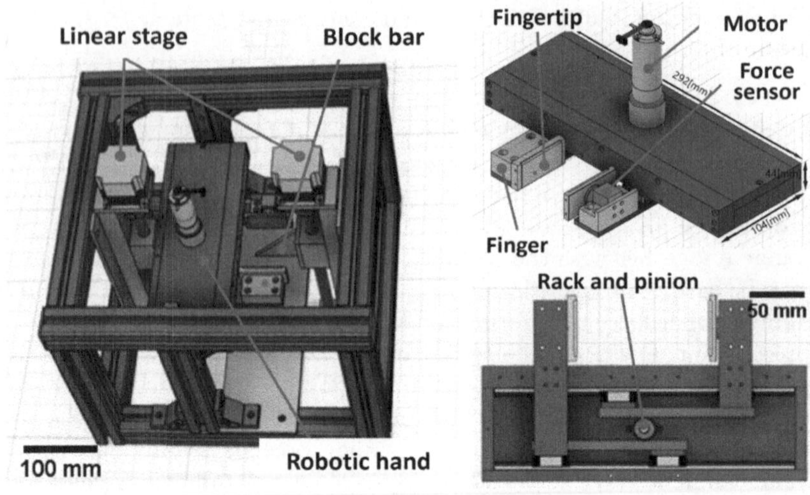

(a) (b)

FIGURE 6.13
Device for measuring viscoelasticity and friction coefficient of foods: (a) overall device design and (b) robotic hand design. Reproduced with permission [52]. Copyright 2021, Elsevier.

Experimental viscoelasticity measurements were conducted on four types of tempura items, as shown in Figure 6.14. The tempuras were purchased from a nearby supermarket, and the tempura shrimp (Figure 6.14a) was cut in half to fit into the measuring device. The masses of tempuras were recorded as 21.1 g (shrimp), 17.2 g (lotus root), 15.2 g (eggplant), and 8.0 g (green pepper), respectively. The measurements were performed immediately after purchase. These tempuras, with irregular shapes and complex properties, were chosen because the property measurements of such food items are influenced more by how they are grasped and measured than by their intrinsic properties.

The measurements were conducted according to the following procedure:

1. The tempura was placed on the working area of the measuring device;
2. The robotic hand was actuated to grasp the tempura at a predetermined velocity using the position control mode;
3. The grasping motion of the robotic hand was stopped when the grasping force measured by the force sensor exceeded a predetermined target force;
4. The position of the robotic hand was maintained for a certain time period to obtain the force relaxation data;

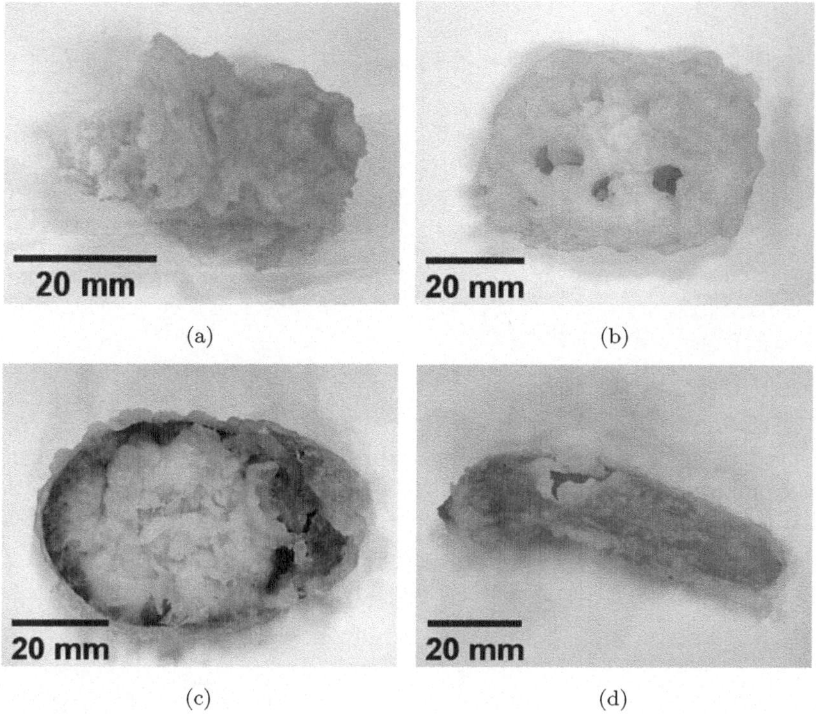

FIGURE 6.14
Food items for viscoelasticity measurements: (a) tempura shrimp, (b) tempura lotus root, (c) tempura eggplant and (d) tempura green pepper. Reproduced with permission [52]. Copyright 2021, Elsevier.

5. The hand was actuated to release the tempura and complete one test trial;
6. The data regarding time, hand displacement, and grasping force were saved.

Examples of measured viscoelastic forces for tempura shrimp under different grasping forces and velocities are shown in Figure 6.15a. It is observed that the same targeted grasping forces generate different values of f_{peak} due to varying grasping velocities. This variation is attributed to the viscosity of the food items. Using the three-step approach described earlier, the parameters related to the viscoelasticity of food can be estimated. An example of the approximated viscoelastic force for a grasping force of 0.5 N and a velocity of 10.0 mm/s is shown in Figure 6.15b. Similar approximation results were obtained for other test conditions.

The estimated viscoelasticity parameters for different tempura items are summarized in Figure 6.16. All three parameters were influenced by both the

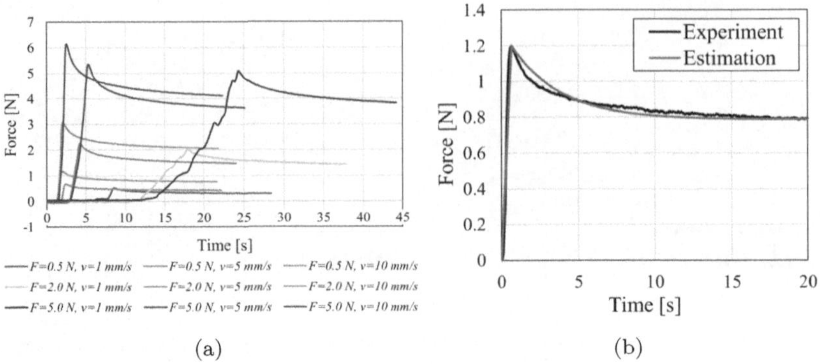

FIGURE 6.15
Measured viscoelastic forces of tempura shrimp (a) and example of fitted force using estimated parameters (b). Reproduced with permission [52]. Copyright 2021, Elsevier.

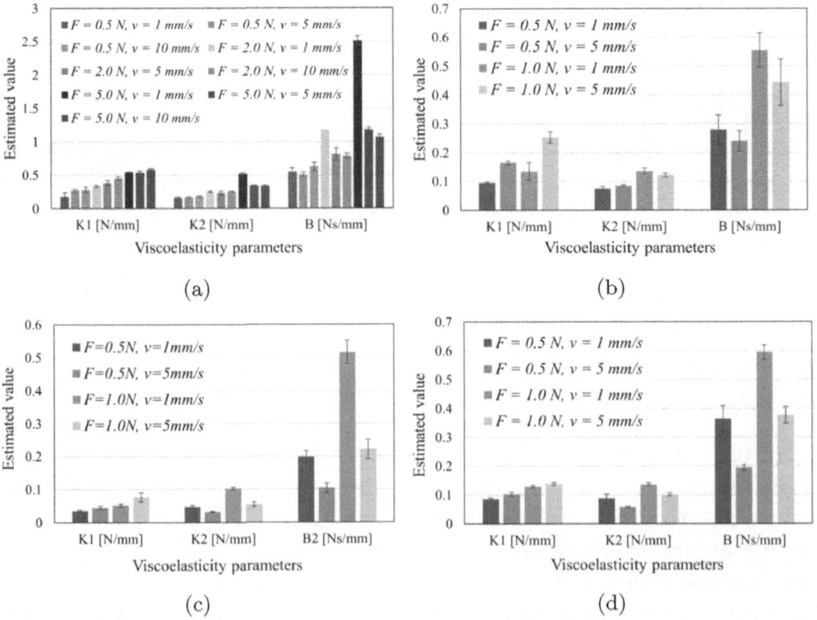

FIGURE 6.16
Viscoelasticity parameters of different food items: (a) tempura shrimp, (b) tempura lotus root, (c) tempura eggplant and (d) tempura green pepper. Reproduced with permission [52]. Copyright 2021, Elsevier.

target force and the grasping velocity. As the target force increased, the values of the estimated parameters generally rose. Conversely, with an increase in the grasping velocity, the parameter K_1 tended to increase, while the parameters K_2 and B exhibited decreasing trends.

6.2.1.3 Measuring elastoplastic properties

For many food items, the deformation caused during grasping does not fully recover, leaving residual deformation that may impact the product's value. To model this residual deformation, elastoplastic models are required. The two most commonly used elastoplastic models are the Maxwell model (Figure 6.3b) and the three-element model (Figure 6.3c), in which an independent damper accounts for the residual deformation. For modeling more complex elastoplastic behaviors, additional fundamental elements can be incorporated, as shown in Figure 6.17. These models can be classified into two categories: serial models (Figure 6.17a) and parallel models (Figure 6.17b). In [51], the constitutive laws and solutions for both model types were theoretically derived, revealing similarities between them. By combining these models with multi-dimensional finite element formulations, the deformation and force behaviors of food materials can be effectively modeled.

Before using elastoplastic models to simulate food items, the parameters involved in these models need to be identified. In [54], an optimization-based approach was proposed to identify the elastoplastic parameters. The optimization process, shown in Figure 6.18, minimizes the difference between the

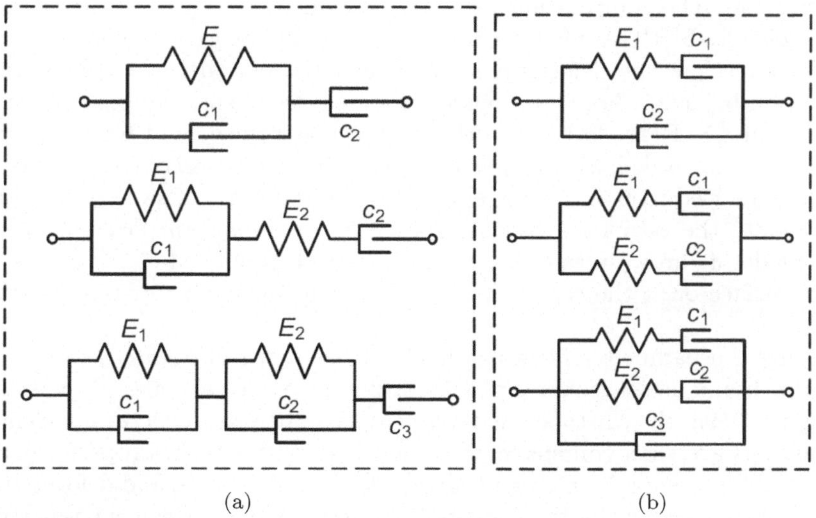

FIGURE 6.17
Constitutive models for elastoplastic food items: (a) serial models and (b) parallel models. Reproduced with permission under the CC-BY 4.0 license [53].

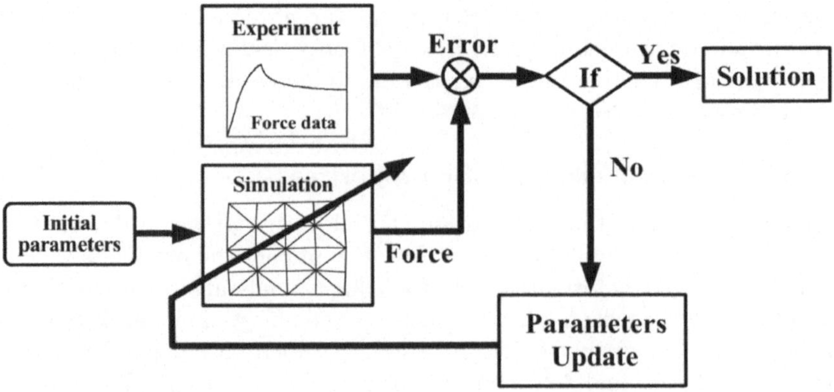

FIGURE 6.18
Optimization process for identifying the elastoplastic parameters. Reproduced with permission [54]. Copyright 2009, IEEE.

experimentally measured force profile and the simulated one, with the parameters being automatically updated during the process.

In [55], the proposed approach was validated using three types of Japanese sweets provided by a sweets company in Kyoto. These sweets, made from a specific ratio of flour, water, and bean powder, are very soft and exhibit typical elastoplastic behaviors. For compression tests, three flat, square-shaped samples, each composed of one of the sweets' materials, were prepared, as shown in Figure 6.19a. Black dots were drawn on the surface to better illustrate the deformation. The compression tests were conducted using a motorized stage (KX1250C-L, SIRIGA SEIKI Co.) at a constant velocity. Before releasing the compression, the deformation was held for a period to record the force relaxation behavior. Force responses were measured using a tactile sensor (I-SCAN100L, NITTA Co.), while the initial, maintained, and final deformed shapes were recorded with a camera (Canon EOS Kiss X2). These measurements were used to estimate the elastoplastic parameters of sweets' materials. To validate the estimated parameters, several non-uniform layered objects, each made of three alternating materials, were compressed either across the entire surface or at the center area, as shown in Figures 6.19b and c, respectively.

Using the parameter identification process shown in Figure 6.18 and the experimentally measured force profiles of the three sweets' materials presented in Figure 6.19a, the elastoplastic parameters can be estimated. Finite element simulations are then conducted to demonstrate the estimation performance. Figure 6.20 presents the fitting results of elastoplastic forces and deformations at various stages for the three sweets' materials. Good agreement was found between the experimental data and the simulation results, validating the parameter identification process and confirming the accuracy of the estimated

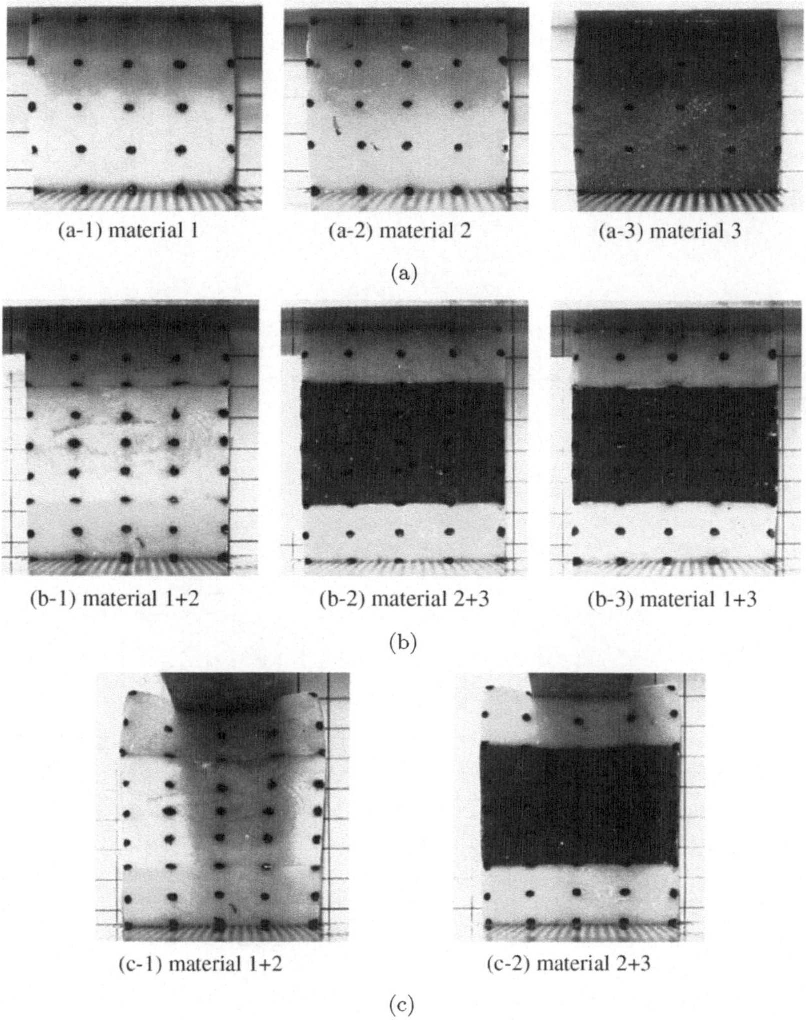

FIGURE 6.19
Sweets' materials and layered structures for property measurements and validations: (a) three types of sweets' materials, (b) layered structures compressed from the entire top surface, and (c) layered structures compressed from the center area of the top surface. Reproduced with permission [55]. Copyright 2011, Elsevier.

elastoplastic parameters in capturing the elastoplastic behaviors of the sweets' materials.

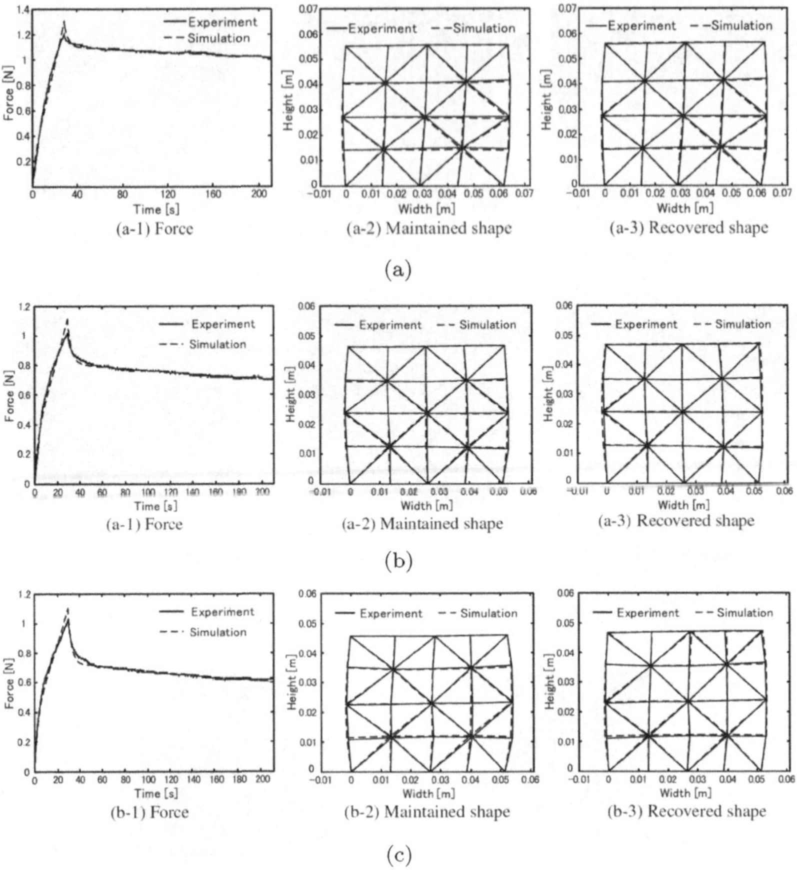

FIGURE 6.20
Simulation results of force and deformation using the identified parameters: (a) sweets' material no. 1, (b) sweets' material no. 2, and (c) sweets' material no. 3. Reproduced with permission [55]. Copyright 2011, Elsevier.

6.2.2 Three-dimensional shape

Shape defines the geometry of a food item. While food items with regular shapes are easier to handle, irregular-shaped ones present more challenges. In food handling tasks, a robotic end-effector can often determine if a food item can be grasped based on its shape. For example, clamping robotic end-effectors are commonly used for food items with two parallel surfaces but are less effective for items without parallel surfaces. Additionally, food items exhibit significant individual variations in shapes and sizes, complicating their handling. Therefore, understanding the 3D shape of food is crucial for designing effective end-effectors. It provides valuable information about the 3D

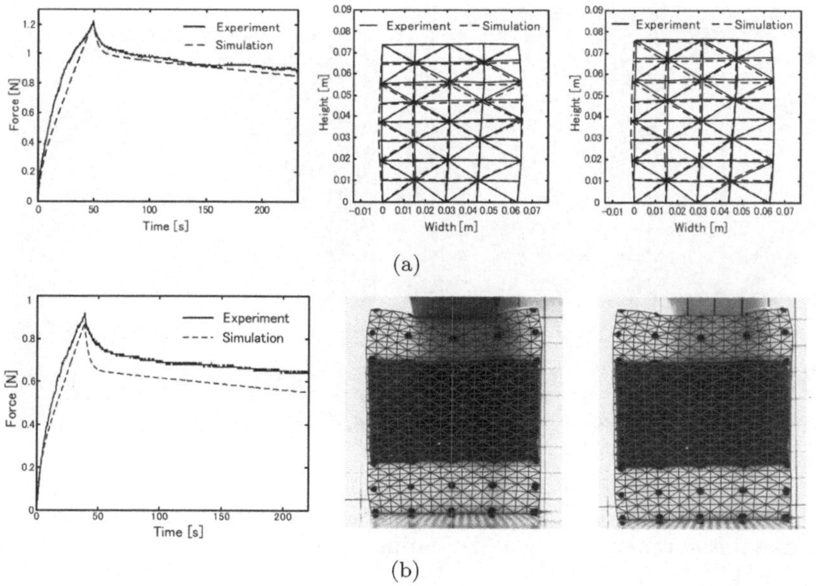

FIGURE 6.21
Validation results of force and deformation for the layered structure: (a) material no. 1 + no. 3 compressed from the entire top surface and (b) material no. 2 + no. 3 compressed from the center area of the top surface. Reproduced with permission [55]. Copyright 2011, Elsevier.

geometry and size of the food, and can also be used in simulations to optimize end-effector designs.

There are various methods for measuring 3D shapes, such as using 3D cameras or 3D scanners. Figure 6.22 illustrates an example of measuring the 3D shape of a green pepper using a fully automated desktop 3D scanner (Artec Micro). This scanner can generate accurate 3D point cloud data, which can then be saved as an STL file for further applications, such as 3D printing. The 3D geometry of the green pepper is shown in Figure 6.22b. Additional examples of 3D scanned food items can be seen in Figure 6.23.

6.2.3 Friction

Friction is a resisting force that prevents two objects from sliding freely against each other. Since the lifting force of most robotic end-effectors relies on friction, it is an important physical property in food handling tasks. In the context of stable grasping, static friction is of particular interest. Static friction can be quantitatively expressed as the maximum static friction coefficient, which is the ratio of the shear force to the normal force during grasping. The maximum

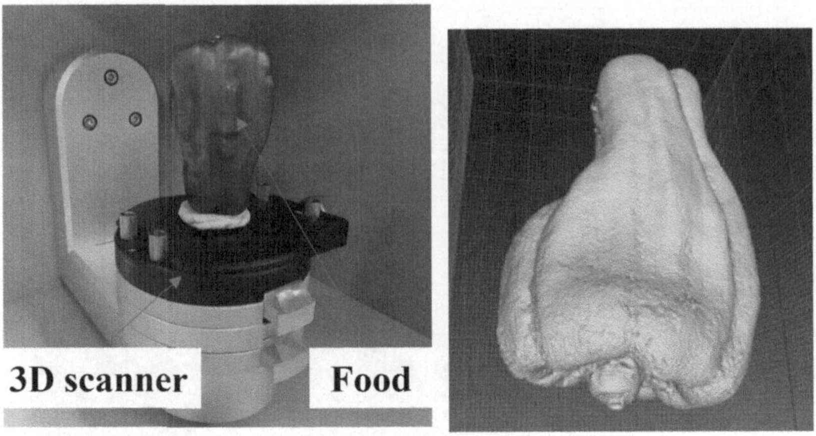

FIGURE 6.22
3D shape measurement using a 3D scanner (a) and the measured 3D shape of a pepper (b).

static friction coefficient of a food item can be used to calculate the required lifting force for a robotic end-effector.

To measure the friction coefficient of food items, one approach involves using a standard friction measurement machine (TL201, Trinity-Lab Inc.,

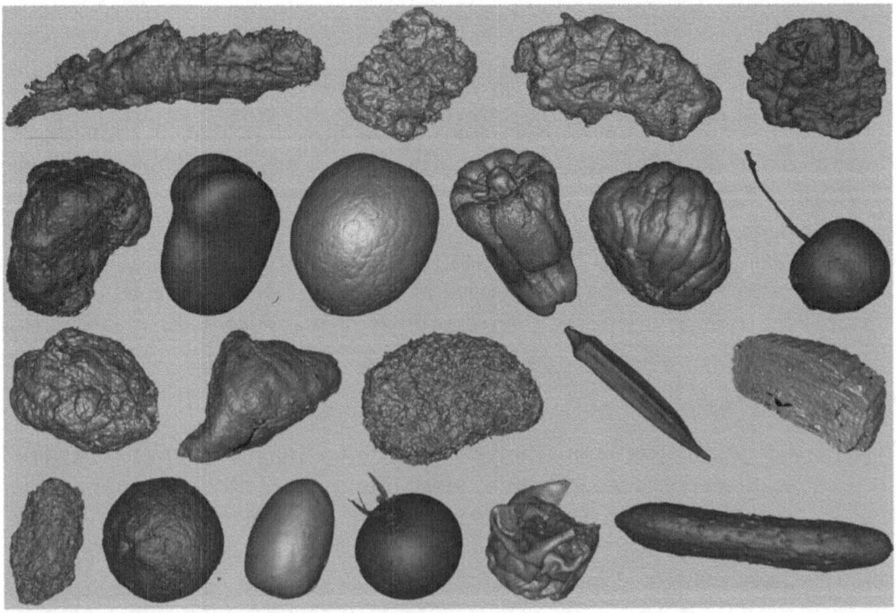

FIGURE 6.23
Examples of measured 3D shape of various food items.

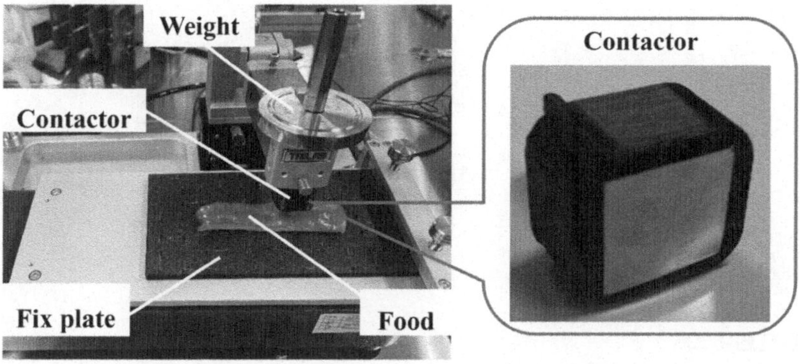

(a)

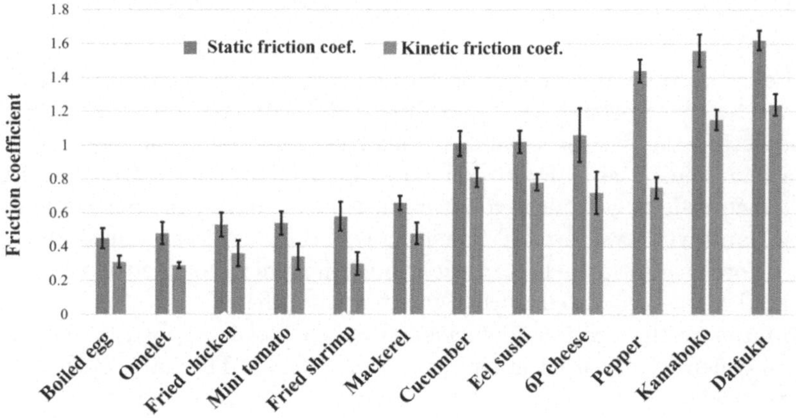

(b)

FIGURE 6.24
Measurement of friction coefficient of various food items: (a) experimental setup and (b) examples of measured friction coefficients of food items shown in Figure 6.11.

Japan), as shown in Figure 6.24. A portion of the food item is cut and placed on a 3D-printed fixture plate. A 3D-printed contactor, covered with a stainless steel plate, is used to make contact with the food item. Dead weight is applied to the contactor to adjust the normal force. The measuring machine is then actuated to translate the fixed plate and food item in a horizontal direction, generating slipping between the contactor and the food item, while the shear force is recorded. The friction coefficients are calculated based on the dead weight and the shear force. Figure 6.24b shows the measured maximum static and kinetic friction coefficients for various food items.

One disadvantage of using a standard friction measuring machine is that a portion of the food item must be prepared and fixed to the plate. This process can introduce measurement errors, particularly for very soft and fragile

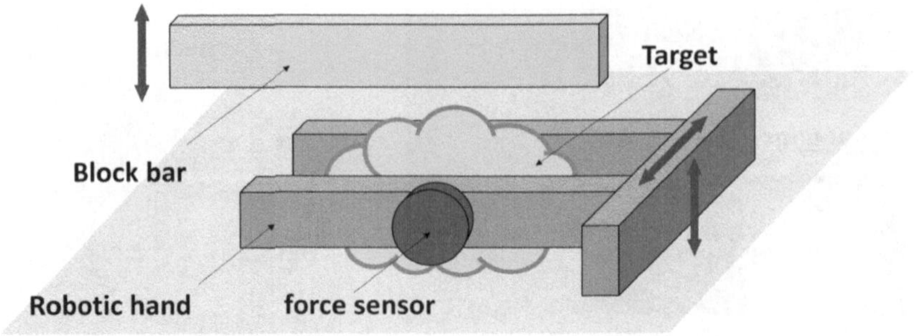

FIGURE 6.25
Principle of measuring friction coefficient using a robotic hand. Reproduced with permission [52]. Copyright 2021, Elsevier.

food items, as they are difficult to fix and the cutting operation may significantly alter their properties. To address this issue, an alternative approach for measuring the friction coefficient involves using a robotic hand equipped with a force sensor, as described in [52]. The principle is illustrated in Figure 6.25, which reflects a typical robotic handling scenario. In this method, the robotic hand with two parallel fingers grasps the food item placed on a platform. A force sensor attached to the robotic fingers measures the normal and tangential forces generated during grasping and lifting.

To measure the friction coefficient, force control is applied to the robotic hand to generate a desired grasping (normal) force. The hand is then translated vertically to lift the food item. A block bar is placed on top of the food item to prevent it from being lifted, inducing slippage between the robotic hand and the food item. The friction coefficient can then be determined using the tangential and normal forces measured by the force sensor. By controlling the applied grasping force, the relationship between the normal force and the friction coefficient can also be explored. The developed measurement device, shown in Figure 6.13, is the same one used for measuring the viscoelasticity of food items.

Using the principle outlined above and the developed measuring device, experiments were conducted on the same tempura items shown in Figure 6.14. The measured normal and shear forces under different predetermined grasping (normal) forces are presented in Figure 6.26a for the tempura green pepper as an example. The jaggedness observed in the curves during sliding is due to vibrations and stick-and-slip behavior during slippage. Friction coefficients are then calculated as the ratio of shear to normal forces. The maximum values of the coefficients for each trial are considered the static friction coefficient, and the results for the four tempura items are summarized in Figure 6.26b.

The results indicate that the tempura items exhibit a friction coefficient of approximately 0.3–0.4 at grasping forces between 0.5 N and 1.0 N, with standard deviations smaller than 0.1. As the normal grasping force increases,

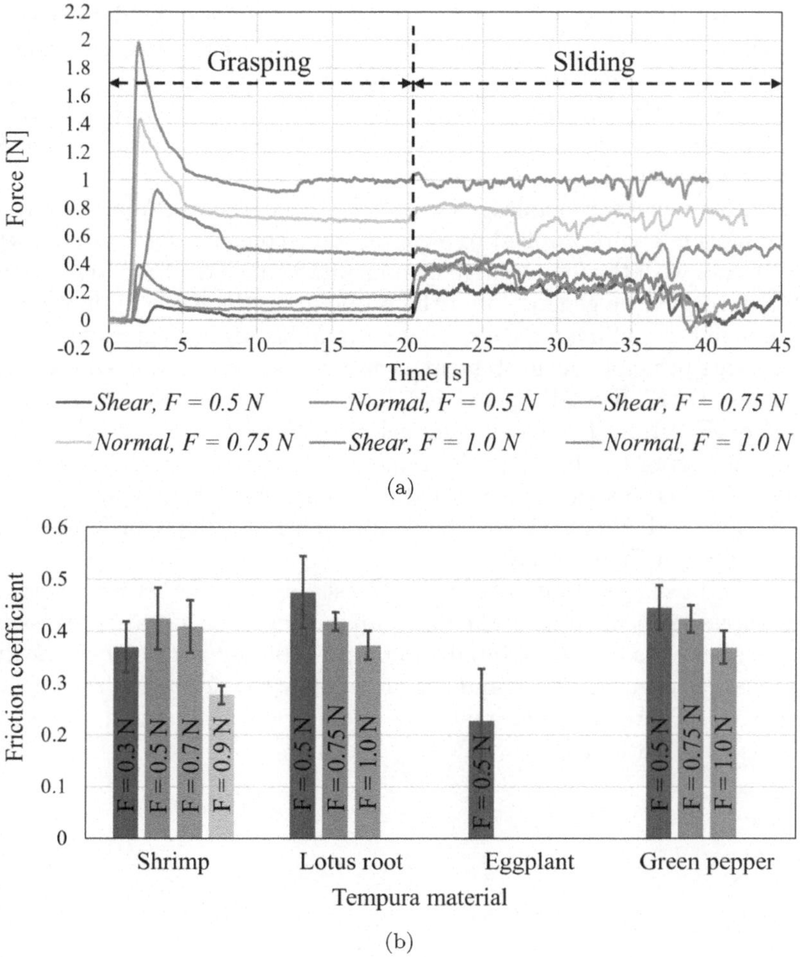

FIGURE 6.26
Measure friction forces (a) and calculated friction coefficients of tempura materials (b). Reproduced with permission [52]. Copyright 2021, Elsevier.

the friction coefficient shows a slight decreasing trend, though the differences are not significant. This suggests that an excessively high grasping force may not only damage fragile food items but could also reduce the frictional force, potentially leading to grasping failure.

6.2.4 Stickiness

Stickiness is a property of food items that causes them to adhere to surfaces, which can affect the performance of robotic hands when releasing food items.

A sticky food item may make it difficult for the robotic hand to place the item accurately at a predetermined position and orientation. Position and orientation errors are crucial for evaluating the performance of robotic hands. However, stickiness has rarely been considered in robotic food handling research. To characterize stickiness, the pulling force upon releasing the food item can be measured.

The measurement principle is illustrated in Figure 6.27a. A push-down and move-up motion can be applied to a probe connected to a force sensor, which makes contact with the food item fixed on a platform. If the food item is sticky, a pulling force will be generated on the probe upon the move-up motion. The magnitude of this pulling force can be used to indicate the degree of stickiness.

Using this principle, measurement experiments were conducted on silicone gel samples (HAPTICS OF WONDER, Taica), a material kit consisting of 12 types of αGEL from Taica's silicone gel collection. The experimental setup is shown in Figure 6.27b, using the same test stands as those for measuring Young's modulus. Instead of using a stainless sphere, a thin, cylinder-shaped indentor was used. The measured stress-strain relationships for four gel samples are shown in Figure 6.27c.

The results suggest that there is hysteresis in the stress-strain curve, and negative stress appears during the move-up motion. This negative stress indicates the stickiness of the material samples. Materials with higher stickiness generate larger negative stress and are more likely to stick to the robotic hand upon release. Therefore, such materials require extra attention when being placed by the robotic hand.

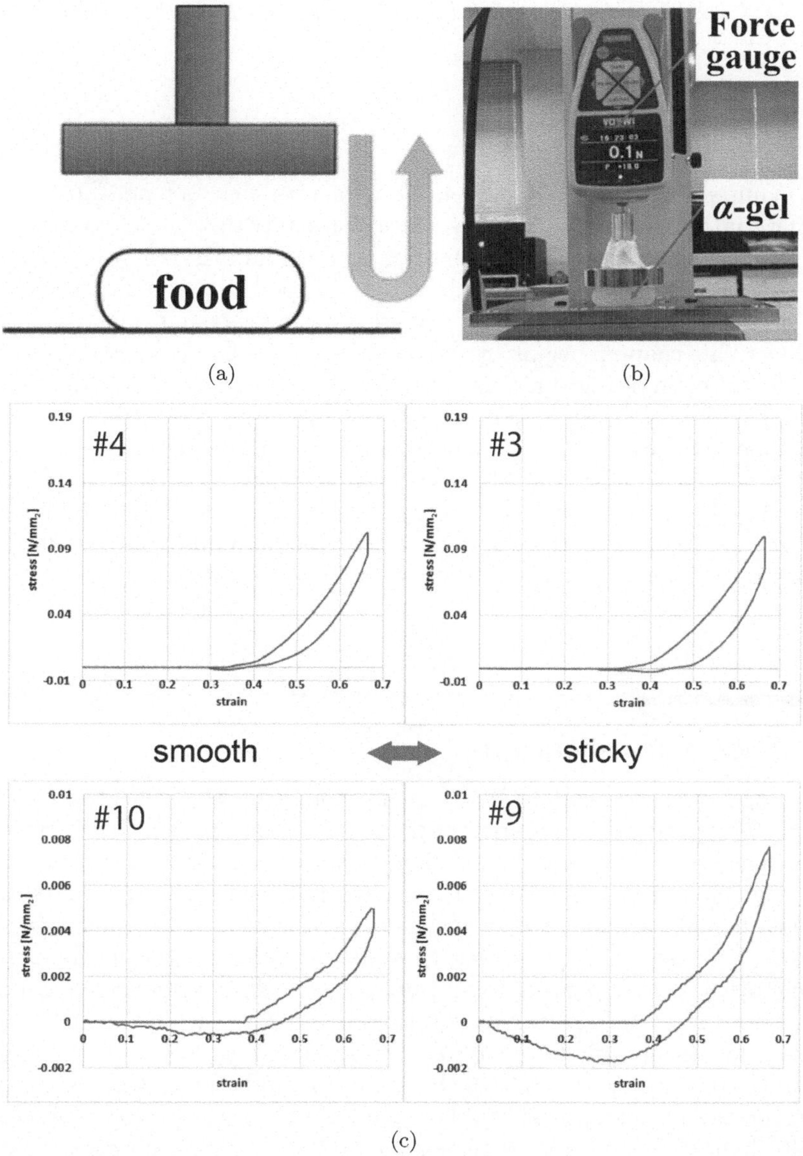

FIGURE 6.27
Measurement of stickiness: (a) measurement principle, (b) experimental setup, and (c) examples of measured sticky force of various gel materials.

7

Food Classification and Sample Fabrication

Food manipulation is heavily influenced by factors such as food shapes, sizes, and material properties, including surface characteristics. Understanding how these properties affect food manipulation is essential. To design robotic end-effectors capable of handling a wide variety of food items, it is necessary to categorize food items based on their physical properties that relate to robotic handling. This chapter begins by introducing a classification system for food items, focusing on physical properties relevant to grasping, such as 3D shapes, elasticity, viscosity, friction, and adhesion. To facilitate the evaluation of different robotic hands, food samples must accurately replicate the physical properties of real food items. Therefore, this chapter also explores the fabrication approaches for creating such food samples. Finally, robotic handling tests on various food items and their corresponding food samples are presented to demonstrate the functionality of these samples.

7.1 Food Classification

Regarding food classification, Wurdemann et al. proposed a method for categorizing food products in the context of food ordering, considering key properties such as symmetry, surface condition, hardness, springiness, and resistance to damage [56]. Similarly, Erzincanli and Sharp classified food products based on attributes like shape, dimensions, surface characteristics, compliance, temperature, and weight [57]. However, these classification systems primarily rely on qualitative descriptions of food properties—such as smooth versus rough surfaces and rigid or hard versus non-rigid or soft compliance—which are insufficient for the specific needs of robotic food handling.

When considering the robotic handling of food items, various physical properties can be categorized. As shown in Figure 7.1, these properties can be broadly divided into surface properties and body properties. Surface properties refer to the external physical characteristics of a food item, such as friction, stickiness, temperature, and humidity. In contrast, body properties describe the internal physical characteristics, including shape, size, weight, elasticity, viscosity, and fragility. The following subsections will define these properties and explain how they influence the robotic handling of food items.

DOI: 10.1201/9781003372431-7

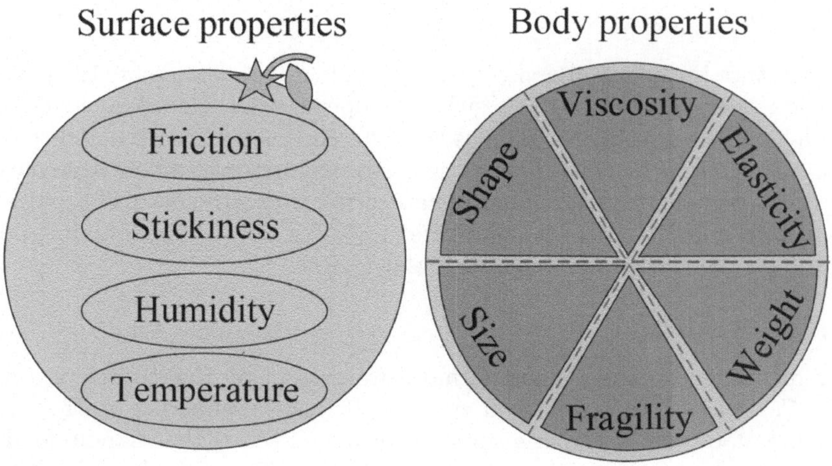

FIGURE 7.1
Food properties related with robotic handling. Reproduced with permission under the CC-BY 4.0 license [58].

7.1.1 Surface properties

7.1.1.1 Friction

Friction is a resisting force that prevents two objects from sliding freely against each other. In robotic food handling, where the lifting force of most end-effectors is generated by friction, this property becomes crucial. For stable grasping, static friction is of particular interest. Therefore, we focus on the static friction coefficient in food handling tasks. The static friction coefficient is quantitatively expressed as the ratio of the maximum friction force to the normal force during grasping. This coefficient can be directly applied to classify food items based on their frictional properties.

7.1.1.2 Stickiness

Stickiness refers to a food item's ability to adhere to a surface, characterized by the attractive force between surfaces. A sticky food surface can cause the food item to stick to the robotic hand, potentially interfering with the placement process. Therefore, stickiness is an important physical property in food handling tasks. To classify different levels of stickiness, we propose evaluating the performance of various end-effectors by measuring the position error after placement. The position error can be quantified, and stickiness levels can be defined as low (0–10 mm), medium (10–20 mm), and high (>20 mm), depending on the application.

7.1.1.3 Temperature

Temperature is a physical quantity that quantitatively represents the perception of hotness or coldness. The surface properties of a food item change depending on its temperature, such as the difference between frozen and thawed meat. Considering that food handling tasks may occur in different scenarios, we can classify temperature into five categories: frozen (below 0°C), cold storage ($0 \leq T < 10$°C), room temperature ($10 \leq T < 30$°C), under boiling point ($30 \leq T < 100$°C), and over boiling point (≥100°C).

7.1.1.4 Humidity

Humidity refers to the amount of moisture present on the surface of a food item. Variations in humidity can affect the friction coefficient, which in turn influences food handling performance. For end-effectors that handle food by freezing its surface, humidity is a critical property. To classify humidity levels, we use "wet" and "dry" as indicators.

7.1.2 Body properties

7.1.2.1 Shape

Shape describes the geometry of the body of a food item. Food items with a regular shape are convenient to handle, but irregularly shaped ones are less convenient to handle. In a food handling task, a robotic end-effector can intuitively determine whether a food item can be grasped based on shape information. For example, a clamping robotic end-effector is typically used for grasping food items with two parallel surfaces; however, it is rarely used for grasping food items without any parallel surfaces. In fact, it is difficult to determine the exact geometrical shape that matches a food item. Thus, a food item is considered to have a certain shape if it is roughly matched. Therefore, we propose to categorize food shapes into three levels: regular, semi-regular, and irregular. The regular level indicates that the food items have one of the following shapes: sphere, cylinder, ellipse, cube, torus, cuboid, or cone. The semi-regular level indicates that the food items have a shape in which most sides and interior angles are different but they have at least two parallel sides. The irregular level indicates that the food items have a shape in which all sides and interior angles are different.

Shape refers to the geometry of a food item. Food items with regular shapes are easier to handle, while irregular shapes are more challenging. In food handling tasks, a robotic end-effector can often determine if a food item can be grasped based on its shape. For instance, clamping end-effectors are commonly used to grasp food items with two parallel surfaces but are less effective for those without parallel surfaces. Since it is difficult to precisely match an end-effector to every food shape, we categorize food shapes into three levels: regular, semi-regular, and irregular.

1. Regular: The food item has a simple, predictable shape, such as a sphere, cylinder, ellipse, cube, torus, cuboid, or cone.
2. Semi-regular: The food item has a shape with mostly different sides and angles, but with at least two parallel sides.
3. Irregular: The food item has a complex shape with no parallel sides or consistent angles.

7.1.2.2 Size

Size refers to the 2D magnitude of a food item's body, which is often determined by the grasping configuration in a food handling task. For example, a robotic end-effector is typically designed to grasp the narrower sides of a fish rather than its wider sides. As such, the length of the grasping side of a food item is used to quantitatively define its size.

7.1.2.3 Weight

Weight refers to the mass or heaviness of a food item. In food handling tasks, the masses of the food items vary. For instance, light food items such as a strawberry may weigh around 20 g, while heavier food items like a watermelon can weigh approximately 2000 g. Generally, most food items weigh up to 200 g. Mass can thus be directly used to categorize food items.

7.1.2.4 Elasticity and viscosity

Elasticity and viscosity describe a food item's ability to store and dissipate energy. These properties are closely coupled and can be quantitatively described by Young's modulus in food handling tasks. The elasticity of a food item influences its handling performance once grasped. Therefore, Young's modulus can be used as an indicator to categorize food items based on their elasticity.

7.1.2.5 Fragility

Fragility represents the resistance of a food item to damage. The grasping force of a robotic hand must be carefully controlled to prevent damaging the food during handling. To classify food fragility, we can use the grasping force F that may cause damage to the food item and divide fragility into three levels: low ($F \geq 5\,\mathrm{N}$), medium ($1 \leq F < 5\,\mathrm{N}$), and high ($\mathrm{F} < 1\,\mathrm{N}$).

7.1.3 Property-based classification

Based on the descriptions of food properties related to robotic handling, we can classify food items as shown in Table 7.1. This classification provides a foundation for further classifying robotic hands according to how they can adapt to these food properties. The classification of robotic hands, based

TABLE 7.1
Food classification based on physical properties related to robotic handling

Surface properties			
Friction coef.	Stickiness	Temperature	Humidity
$0 < f \leq 0.5$	Low	$< 0°C$	Wet
$0.5 < f \leq 1.0$	Medium	$0 \leq T < 10°C$	Dry
$1.0 < f \leq 1.5$	High	$10 \leq T < 30°C$	
$1.5 < f \leq 2.0$		$30 \leq T < 100°C$	
≥ 2.0		$\geq 100°C$	
Body properties			
Shape	Weight	Elasticity	Fragility
Regular	$0 < m \leq 50\,g$	$0 < E \leq 10\,kPa$	Low
Semi-regular	$50 < m \leq 100\,g$	$10 < E \leq 10^2\,kPa$	Medium
Irregular	$100 < m \leq 150\,g$	$10^2 < E \leq 10^3\,kPa$	High
	$150 < m \leq 200\,g$	$10^3 < E \leq 10^4\,kPa$	
	$\geq 200\,g$	$\geq 10^4\,kPa$	

on their adaptability to different food properties, is introduced in the next chapter.

7.2 Food Sample Fabrication

For experiments involving various robotic end-effectors, actual food items can be used for testing at the final stage. However, at the initial stages, food samples can be very helpful. Actual food items are often difficult to store, and their properties can change over time. In particular, when comparing the performance of different robotic end-effectors, food samples provide consistent standards, ensuring fair and valid comparisons. While many commercially available food samples, such as those shown in Figure 7.2, resemble real food items in appearance, they often fail to replicate important handling-related properties, such as softness, stickiness, and friction. In this chapter, we introduce several examples of food sample fabrication designed specifically for robotic food handling.

7.2.1 Sandwich sample

A sandwich is a popular food item commonly found in convenience stores and supermarkets. A typical sandwich, as illustrated in Figure 7.3a, consists of triangular-shaped bread slices, sliced ham, and sliced lettuce. Based on

FIGURE 7.2
Examples of commercial food samples.

this sketch, the fabricated sandwich sample is shown in Figure 7.3b, which is packaged in a commercially available sandwich bag.

To fabricate the sandwich, all components were produced separately and then assembled. The triangular bread pieces were fabricated from sponge material by Mitsui Chemicals [59]. The ham and lettuce were created by SATO Sample [60] using polymer materials through a molding process. These

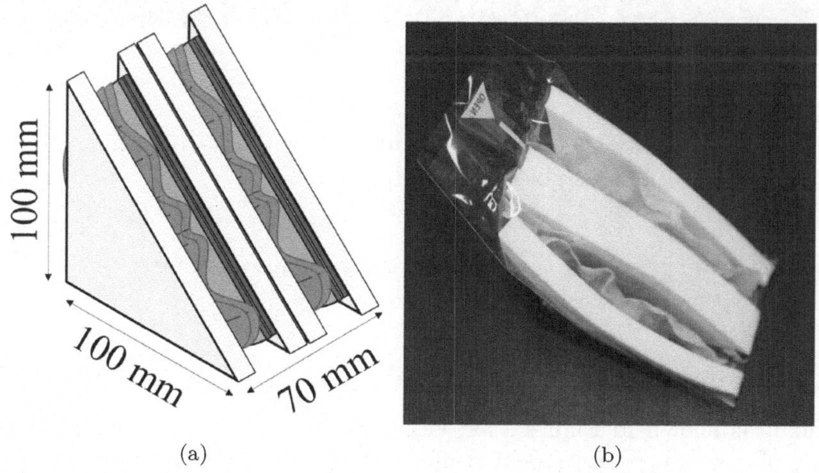

FIGURE 7.3
Sandwich sketch (a) and the fabricated sandwich sample (b).

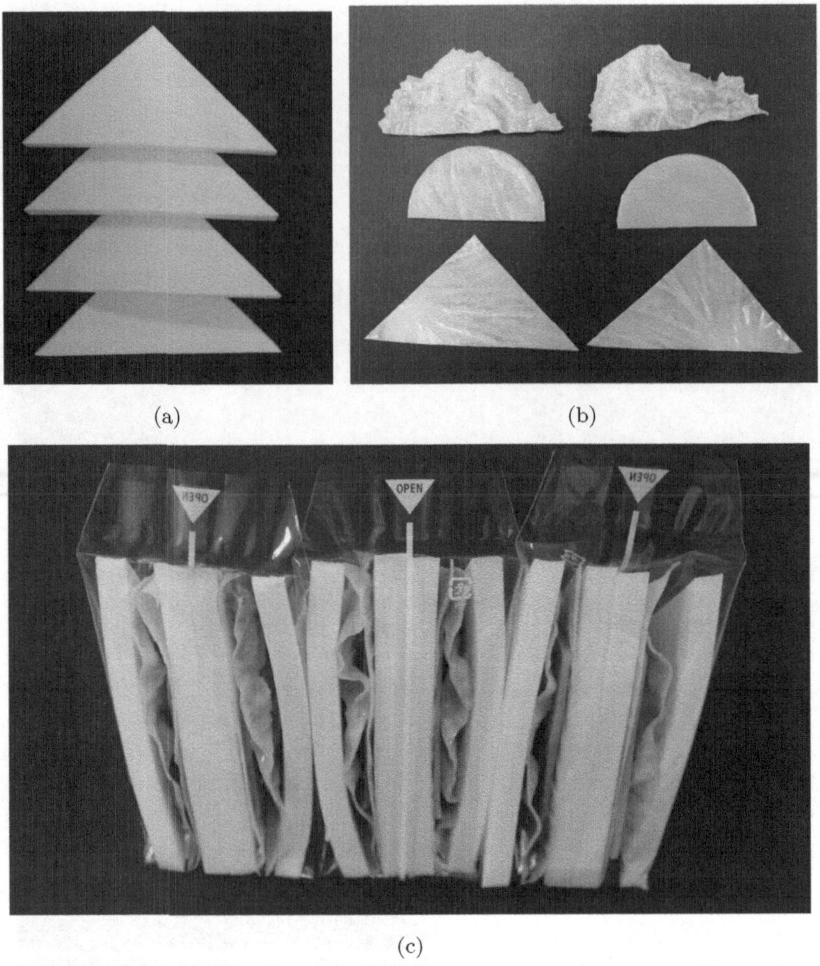

FIGURE 7.4
Fabrication of sandwich sample: (a) bread piece samples, (b) ham and lettuce samples, and (c) the assembled sandwich samples.

components were then assembled and packaged into commercially available sandwich bags, as shown in Figure 7.4c.

7.2.2 Daifuku sample

Daifuku, as shown in Figure 7.5a, is a traditional Japanese confection consisting of a small, round mochi stuffed with a sweet filling, most commonly anko—sweetened red bean paste made from azuki beans. Daifuku is a popular wagashi in Japan and is often served with green tea. To fabricate the daifuku

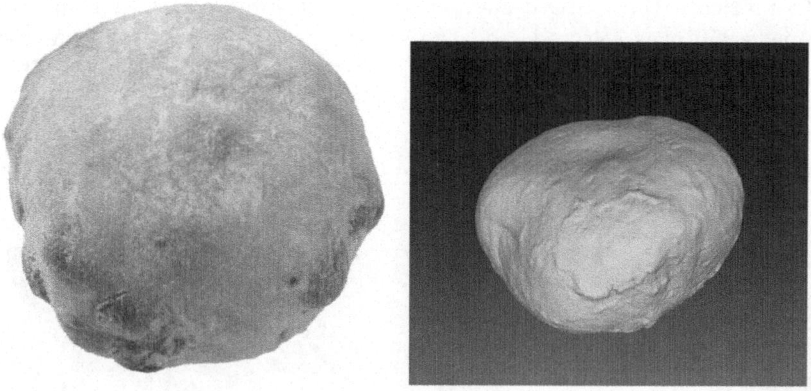

(a) (b)

FIGURE 7.5
Daifuku (a) and its 3D-scanned data (b).

sample, we first measured the 3D geometry of the daifuku using a desktop 3D scanner (Artec Micro, Artec 3D). The measured 3D geometry is shown in Figure 7.5b and can be saved as an STL (STereoLithography) file for further use.

Daifuku is a very soft food, and its softness is crucial for robotic handling. To replicate its softness and surface properties, we measured the stress-strain relationship and friction coefficient, with the results shown in Figure 7.6. These measurements were conducted using the experimental setups illustrated in Figures 6.24 and 6.27, respectively. Based on these measurements, suitable materials for fabricating the daifuku sample were selected.

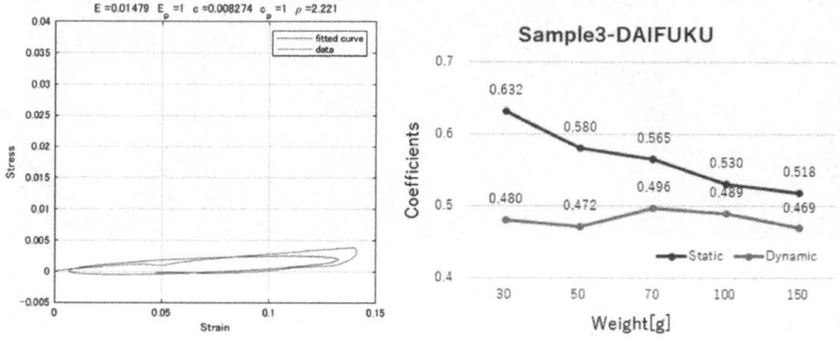

(a) (b)

FIGURE 7.6
Physical parameter measurement of daifuku: (a) stress-strain relationship and (b) friction coefficient.

To fabricate the daifuku sample, super-soft urethane resin from Mitsui Chemicals [59] was selected, and the sample was fabricated using a casting process. An example of the fabricated sample is shown in Figure 7.7a. For packaging the fabricated samples, a professional packaging machine (FUJI WRAPPER FW3301II, FUJI Machinery) was used, as illustrated in the packaging scenario in Figure 7.7b. The final packaged daifuku samples are shown in Figure 7.7c.

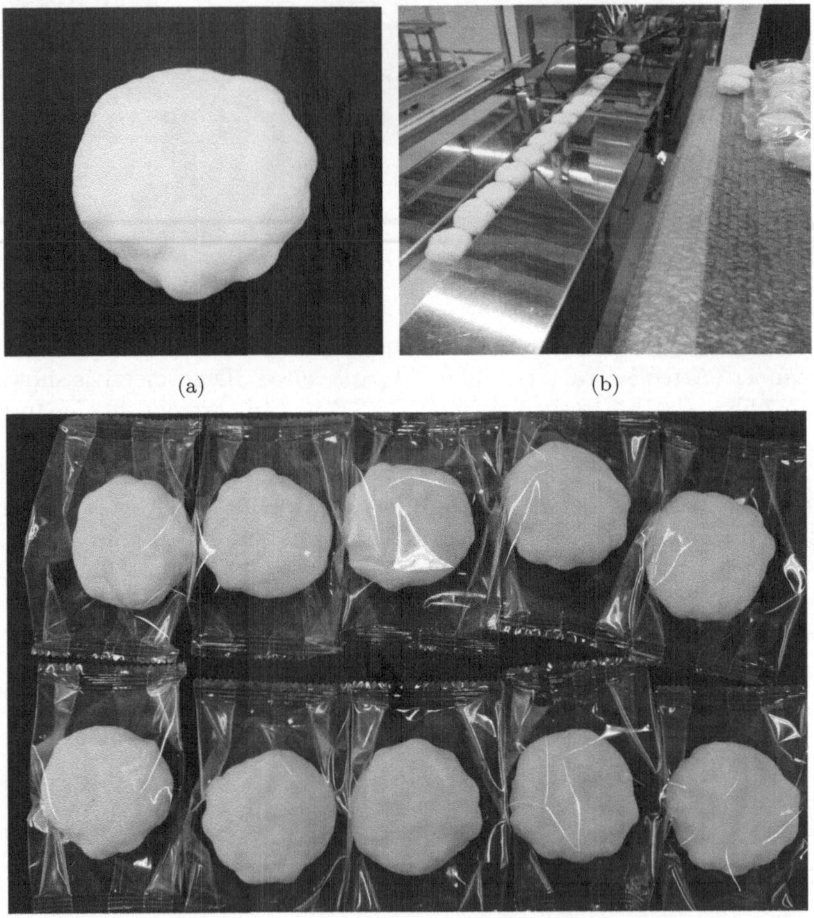

(a) (b) (c)

FIGURE 7.7
Fabricated and packaged daifuku samples: (a) fabricated daifuku sample, (b) packaging process, and (c) the packaged daifuku samples.

7.2.3 Croquette sample

A croquette is a deep-fried food originating in the Mediterranean region, consisting of a thick binder combined with a filling, which is then breaded. It is commonly served as a side dish, snack, or fast food worldwide [61]. A beef croquette is shown in Figure 7.8a as an example. The 3D scanned data of the croquette is shown in Figure 7.8b, and the STL file can be used for fabricating the croquette sample. The stress-strain relationship and friction coefficient were also measured, with the results shown in Figures 7.8c and d, respectively. As observed in Figure 7.8c, hysteresis is clearly present in the real food. From Figure 7.8d, it is evident that the friction coefficient of the croquette is less than 0.3.

The croquette sample was fabricated using silicone and a casting process by EBISU [62]. One example of the fabricated croquette is shown in Figure 7.9a. Its appearance closely resembles the actual food item, but the surface condition does not replicate the actual croquette, which is typically coated with breadcrumbs. To mimic the surface condition, breadcrumbs were used to coat the fabricated croquette. The breadcrumbs are commercially available food materials, and the packaged breadcrumbs are shown in Figure 7.9b,

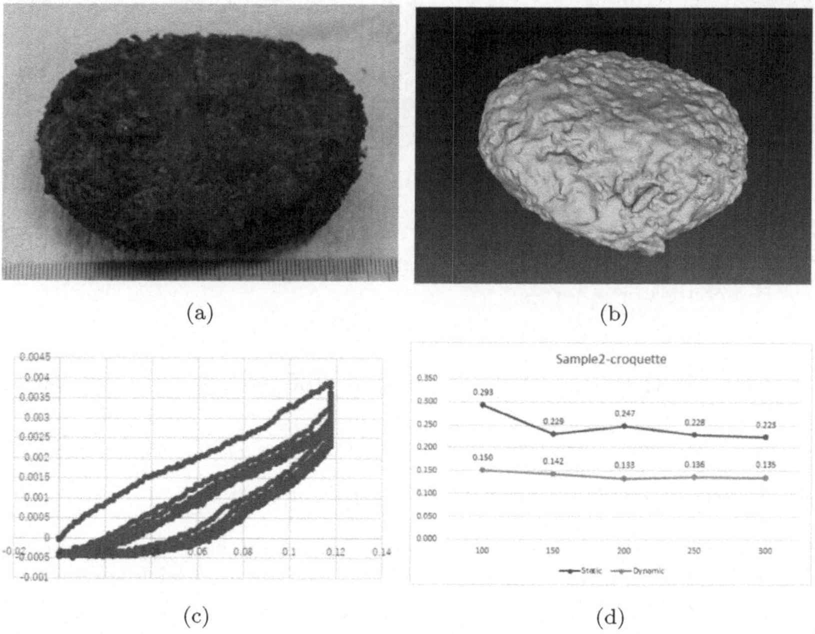

FIGURE 7.8
Croquette measurements: (a) real croquette, (b) 3D scanned croquette, (c) stress-strain relationship of croquette, and (d) friction coefficient of croquette.

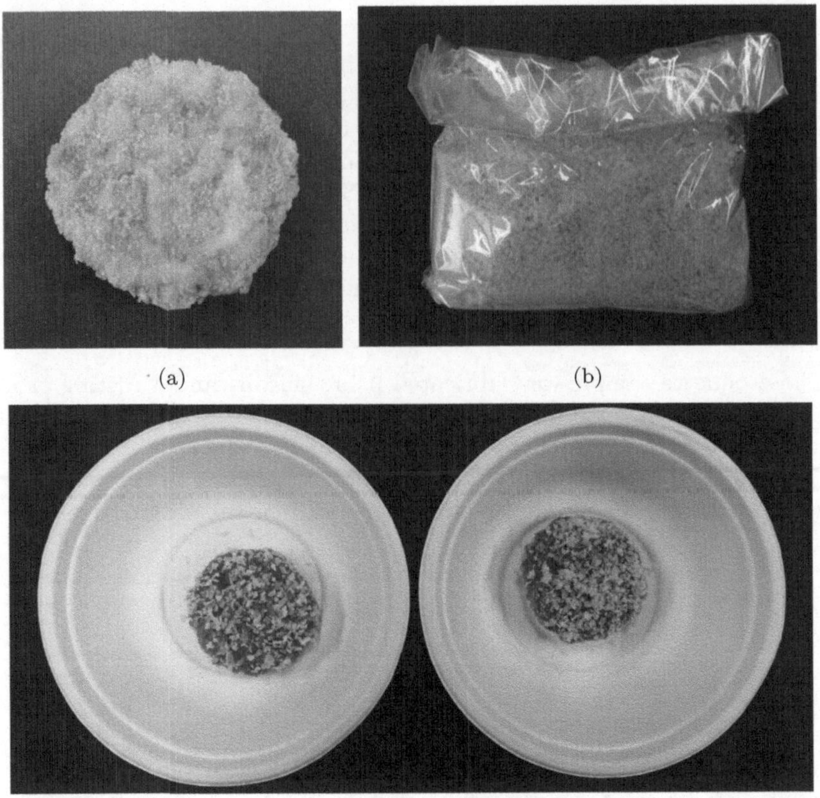

(a) (b) (c)

FIGURE 7.9
Fabricated croquette sample: (a) the sample example, (b) the coating powder, and (c) the coated croquettes.

provided by SATO Sample [60]. The coated croquette samples are shown in Figure 7.9c.

7.2.4 Sample of Japanese hamburger steak

Hamburger steak is a popular dish in Japan, made from ground meat mixed with finely chopped onion, egg, and breadcrumbs, flavored with various spices, and shaped into a flat, circular patty approximately 4 cm thick and 10–15 cm in diameter. Many restaurants specialize in different styles of hamburger steak [63]. A frozen hamburger steak, shown in Figure 7.10a, was used as an example for 3D scanning, as the cooked version is too fragile to be properly fixed on the scanner. The 3D scanned model is shown in Figure 7.10b. The stress-strain relationship was measured using the cooked hamburger steak,

FIGURE 7.10
Measurements of Japanese hamburger steak: (a) real hamburger steak, (b) 3D scanned hamburger steak model, (c) stress-strain relationship of hamburger steak, and (d) friction coefficient of hamburger steak.

with results shown in Figure 7.10c. Compared to croquette and daifuku, the hamburger steak showed a smaller stress, indicating that it is softer. Hysteresis was also observed in the hamburger steak. The friction coefficient measurements, presented in Figure 7.10d, revealed that the static friction coefficient is relatively large compared to daifuku and croquette, which may be attributed to the sauce on the hamburger's surface.

To fabricate samples of Japanese hamburger steak, 3D printing technology was employed due to its wide availability. The fabricated hamburger steak samples are shown in Figure 7.11. For the fabrication process, a popular FDM (Fused Deposition Modeling) 3D printer (Prusa MK4) was used in conjunction with Hotty Polymer HP filament [64], a flexible material. To replicate the softness of the hamburger steak, the print's infill density was set to 15%.

FIGURE 7.11
Examples of fabricated hamburger steak samples.

7.3 Grasping Tests

Grasping tests were conducted using both fabricated food samples and real food items with different robotic systems to compare their performance and assess the differences in handling.

7.3.1 Tests on grasping sandwich

Experimental tests on grasping a sandwich were conducted using the commercial food-handling robot (Foodly, RT Corp.) [65]. Figure 7.12 presents snapshots of the experiment with both the real food item and a fabricated sample. The robot identified the food item in the container, after which the gripper was controlled to grasp the sandwich and place it onto a tray provided by the conveyor. The robot successfully recognized, grasped, and placed both the real food item and the sample.

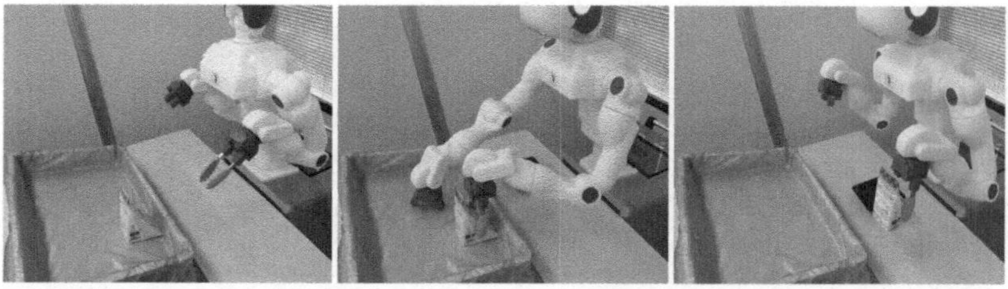

(a)

(b)

FIGURE 7.12
Grasping tests on sandwich using robot Foodly: (a) handling real sandwich and (b) handling sandwich sample.

7.3.2 Tests on grasping daifuku

Grasping tests on daifuku were performed using the mGrip gripper (mGrip, Soft Robotics Inc.) [66]. The mGrip gripper, a pneumatically driven soft gripper, is designed to handle various types of objects. It was mounted on a 4-DOF SCARA robot manipulator (HSR-065, DENSO WAVE). A 3D camera (RealSense D435i, Intel) was used to recognize the target food item, and the robot manipulator was then controlled to execute a pick-and-place action. Figure 7.13 shows experimental snapshots of handling both the real food item and a fabricated sample. The results demonstrated similar performance of the gripper in handling both the real food item and the sample.

7.3.3 Tests on grasping hamburger steak

Handling tests on hamburger steaks were also conducted using the same robotic system as in the daifuku tests. Experimental snapshots are shown in Figure 7.14. The mGrip gripper successfully handled both the real hamburger steak and its fabricated sample, though the success rates differed. The gripper achieved a 100% success rate with the hamburger sample but only reached a 60% success rate with the real food item. This difference is attributed to the sample's lower mass compared to the actual hamburger steak due to a low infill setting (15%) during 3D printing. To more accurately replicate the mass of a real hamburger, dead weight could be embedded inside the sample during the 3D printing process.

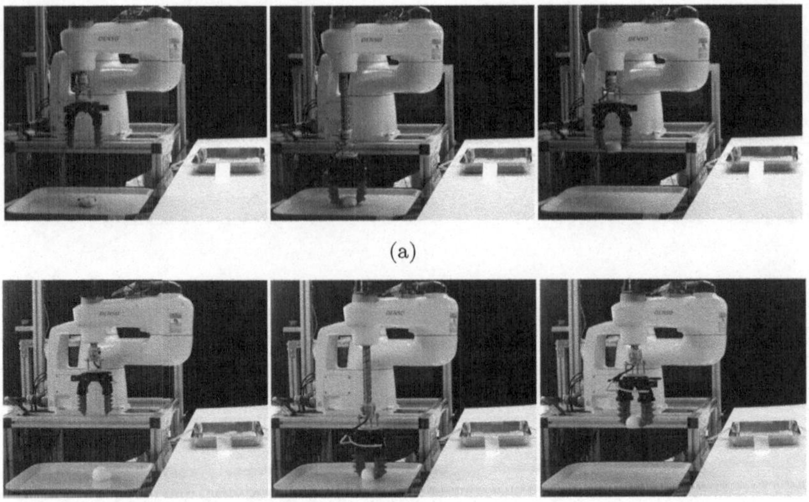

(a) (b)

FIGURE 7.13
Grasping tests on daifuku using mGrip gripper: (a) handling real daifuku and (b) handling daifuku sample.

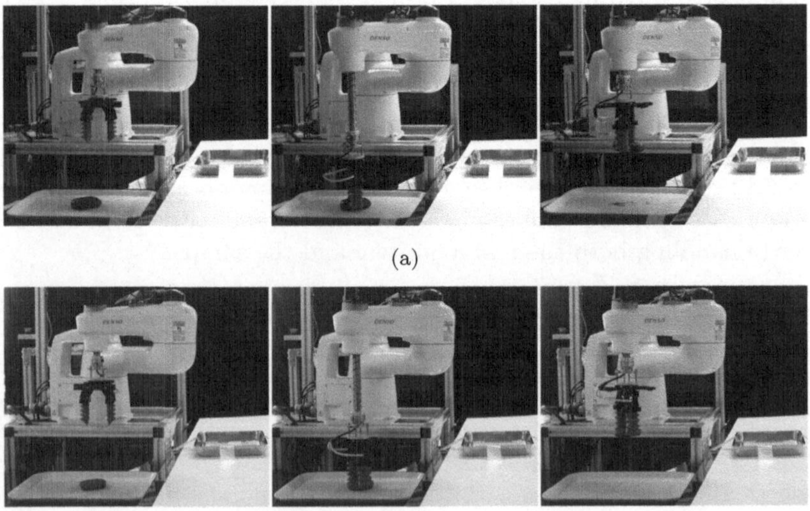

(a) (b)

FIGURE 7.14
Grasping tests on hamburger using mGrip gripper: (a) handling real hamburger and (b) handling hamburger sample.

8

Robotic Hand Classification and Evaluation

To select an appropriate robotic hand for handling specific groups of food items, we must classify robotic hands based on their handling capabilities for the target food groups, following an initial classification of the food items. This classification should be validated through experimental testing and evaluations. For efficient classification and evaluation, a systematic approach is required. This chapter introduces such an approach and demonstrates its effectiveness in helping users select suitable robotic hands for handling food items.

8.1 Classification of Robotic Hand

In their comprehensive review, Fantoni et al. analyzed grasping devices designed for automated production processes [6]. The authors summarized key grasping principles and examined robotic hands across diverse applications, including mechanical, electronic, and micro-assembly, as well as in the food industry, logistics, and integrated grasping and processing tasks. Recently, soft robotic hands have gained considerable attention, particularly for food industry applications. Soft grippers offer advantages such as being lightweight, easily controllable, adaptable, and gentle—characteristics well-suited for food handling. Consequently, soft robotic hands have become the most frequently commercialized soft robots. Examples include the mGrip grippers [66], Soft Robotics flexible grippers [67], modular soft grippers [68], vacuum-driven soft grippers [69], and Rochu soft grippers [70]. Alongside these commercialized soft hands, numerous underdeveloped prototypes tailored for food handling tasks have emerged, including those discussed in this book. A detailed review of the principles and applications of soft robotic hands can be found in [14].

To classify robotic hands based on grasping principles, we adopt the categorization method proposed by Wang et al. [58]. As illustrated in Figure 8.1, robotic hands are divided into seven categories according to the contact position(s) involved in grasping a food item.

DOI: 10.1201/9781003372431-8

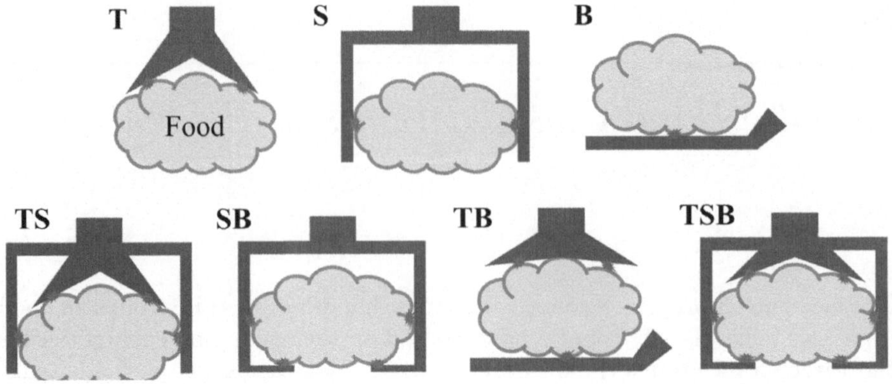

FIGURE 8.1
Robotic hand classification based on grasping principle. Reproduced with permission under the CC-BY 4.0 license [58].

1. The **T-type** indicates that the robotic hand handles a food item only at the top surface, such as the suction cups, Bernoulli principle-based grippers, adhesive hands, and needle grippers.
2. The **S-type** denotes that only the side surfaces of the food item are used during grasping and many two- or multi-fingered robotic hands belong to this category.
3. The **B-type** indicates that the robotic hand will be inserted under the food item for stable handling, such as the SWITL hand commercialized by FURUKAWA KIKO [71].
4. The **TS-type** is a combination of **T-** and **S-type**, and robotic hands belonging to this type handle food items using top and side surfaces simultaneously, such as the gripper commercialized by RightHand Robotics, Inc., which combines a suction cup and a three-fingered gripper [72].
5. The **SB-type** is a combination of **S-** and **B-type**, and the robotic hands handle the food item using bottom and side surfaces simultaneously, such as the ones proposed in [73, 74].
6. The **TB-type** is a combination of **T-** and **B-type**, and the robotic hands of this type handle food items using their top and bottom surfaces.
7. The **TSB-type** is a combination of **T-**, **S-**, and **B-type**, and the robotic hands belonging to this category tend to envelop the food item and use all available surfaces to stabilize the grasping action, such as the multi-functional gripper proposed in [75].

8.2 Evaluation Approach of Robotic Hand

To assess different robotic hands for handling various food items, Wang et al. [58] proposed a scoring method based on the physical properties of food items relevant to robotic handling. To enhance the clarity of evaluation results, a radar chart visualization approach was also introduced.

8.2.1 Scoring method

In the proposed scoring method, each food property is classified into levels to assign a score based on the robotic hand's adaptability to that property. According to the food classification, seven properties were considered in scoring: shape, size, weight, elasticity, fragility, friction, and stickiness. Detailed descriptions of the scoring method are provided below.

8.2.1.1 Shape adaptability

As shown in Table 7.1, food item shapes are categorized into three levels: regular, semi-regular, and irregular. Shape adaptability is defined as the highest shape level that the robotic hand can successfully grasp, with scores assigned as 1, 2, and 3 points for regular, semi-regular, and irregular shapes, respectively. If the robotic hand fails to handle a food item due to its shape, a score of 0 is assigned.

8.2.1.2 Size adaptability

Size is categorized into three levels: small (0–50 mm), normal (50–100 mm), and large (over 100 mm), representing the 2D dimensions of the food item. Size adaptability is scored as 1, 2, and 3 points for the ability to grasp small, normal, and large items, respectively. If the robotic hand cannot grasp the food item due to its size, a score of 0 is assigned.

8.2.1.3 Weight adaptability

Weight is categorized into three levels: light (0–50 g), normal (50–200 g), and heavy (over 200 g). Weight adaptability is scored as 1, 2, and 3 points for the ability to grasp light, normal, and heavy items, respectively. If the robotic hand cannot grasp the food item due to its weight, a score of 0 is assigned.

8.2.1.4 Friction adaptability

Friction is categorized into three levels based on the maximum static friction coefficient: low (0–0.50), medium (0.50–1.00), and high (over 1.00). Unlike shape, size, and weight adaptability, friction adaptability is scored as 3, 2, and 1 points for the ability to handle low, medium, and high friction levels, respectively, reflecting the ability to grasp slippery items. If the robotic hand cannot grasp the food item due to friction, a score of 0 is assigned.

8.2.1.5 Elasticity adaptability

Elasticity is categorized into three levels: low (0–100 kPa), medium (100–500 kPa), and high (over 500 kPa), corresponding to Young's modulus values. Elasticity adaptability is scored as 3, 2, and 1 points for the ability to handle low, medium, and high elasticity levels, respectively, to assess the capacity to grasp soft items. If the robotic hand cannot grasp the food item due to its softness, a score of 0 is assigned.

8.2.1.6 Stickiness adaptability

During a pick-and-place operation, the stickiness of food items can aid in grasping but may also adversely affect the positional accuracy of placement. As shown in Table 7.1, stickiness is categorized into three levels: small, medium, and large, corresponding to positional errors of 0–10 mm, 10–20 mm, and over 20 mm, respectively. Stickiness adaptability is scored as 3, 2, and 1 points for small, medium, and large levels, respectively. A score of 0 is assigned if the robotic hand fails to grasp the food item for other reasons. To measure positional accuracy during experiments, a paper sheet with concentric circles at predefined intervals can be placed, with the center representing the desired placement position. Positional error can then be measured following placement.

8.2.1.7 Fragility adaptability

Fragility is a critical measure of a food item's resistance to damage during and after placement, including rupture, collapse, and plastic deformation. In pick-and-place operations, fragility is closely related to the grasping force exerted by the robotic hand. Since it is challenging to measure absolute grasping forces for soft robotic hands, we used control input (air pressure for pneumatic hands) as a criterion to evaluate fragility adaptability. In the experiments, three air pressure levels were used: 60%, 80%, and 100% of the recommended pressures for positive-pressure-driven soft hands, and −40 kPa, −60 kPa, and −80 kPa for vacuum-driven soft hands. Fragility adaptability was thus categorized into low, medium, and high levels corresponding to these pressure settings. A high level indicates that the food item can be handled without damage even at the highest air pressure. Consequently, fragility adaptability is scored as 1, 2, and 3 points for low, medium, and high levels, respectively, with a score of 0 assigned if the food item is damaged at the lowest air pressure.

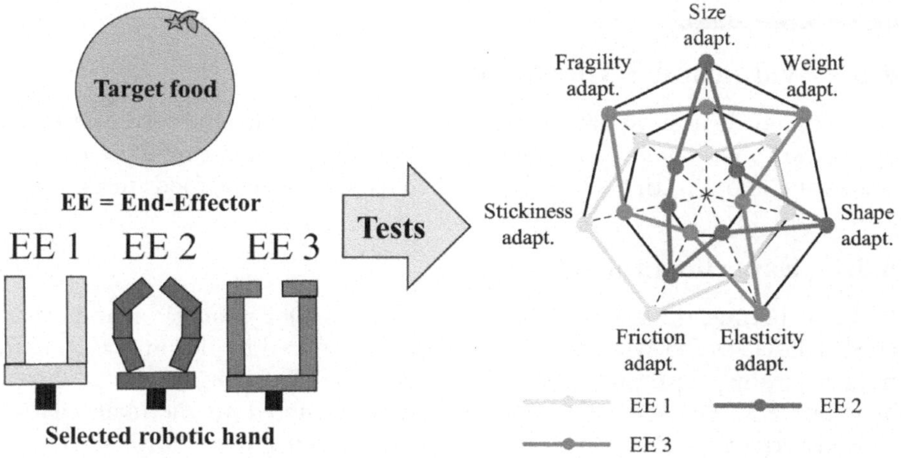

FIGURE 8.2
Performance comparison and visualization of different robotic hands. Reproduced with permission under the CC-BY 4.0 license [58].

As with stickiness adaptability, a score of 0 is also given if the robotic hand cannot grasp the food item for other reasons.

8.2.2 Visualization approach

Using the performance scores described above, a radar chart can be plotted to visualize and compare the capabilities of different robotic hands, as shown in Figure 8.2. The radar chart has seven axes, each representing the adaptability score of a robotic hand for a specific food property. The overall performance of each robotic hand is depicted as a polygon in the radar chart. If a vertex of the polygon is located at the center point, it indicates a score of 0, suggesting that the robotic hand was unable to handle the food item due to one particular property. For example, in Figure 8.2, three robotic hands were evaluated for handling a target food item, with results displayed in the radar chart. The three colored lines correspond to the evaluation results of each robotic hand. The robotic hand EE3 scored [2, 3, 1, 3, 1, 2, and 3], indicating it can grasp a food item with a normal size (50–100 mm), heavy weight (over 200 g), regular shape, medium elasticity (100–500 kPa), low friction (0–0.5), medium stickiness (placement error of 10–20 mm), and high fragility. This radar-chart-based visualization approach enables not only the comparison of different robotic hands for handling the same food item but also the evaluation of a single robotic hand's performance across various food items.

8.3 Evaluation Experiments

Using the evaluation approach outlined above, experiments were conducted on several robotic hands to demonstrate how this method can assist users in comparing and selecting suitable robotic hands for specific food items.

8.3.1 Experimental setup

In the experiments, a four degrees of freedom (4DoF) robotic manipulator (HSR-065A1, Denso Wave Inc., Aichi, Japan) was used to pick up a food item from a picking plate and place it on a designated placing plate, as shown in Figure 8.3. The selected robotic hand was mounted on the manipulator. A measurement scale was positioned on the placing plate, with its center marking the desired placement position of the food item. The robotic manipulator was equipped with a 3D camera (RealSense D435i, Intel, USA) to detect the position and orientation of the food item. To actuate the pneumatic robotic hands, an air compressor (SLP-15EFD, Anest Iwata Corp., Japan) and pressure regulators (ITV2030, SMC Co., Japan) provided the required air pressures. Solenoid valves controlled airflow and were switched through the manipulator's IO interface, while vacuum pressure was generated with an air ejector (VUH07-44A, PISCO Co., Ltd., Japan). The entire robotic system was operated using ROS 2.

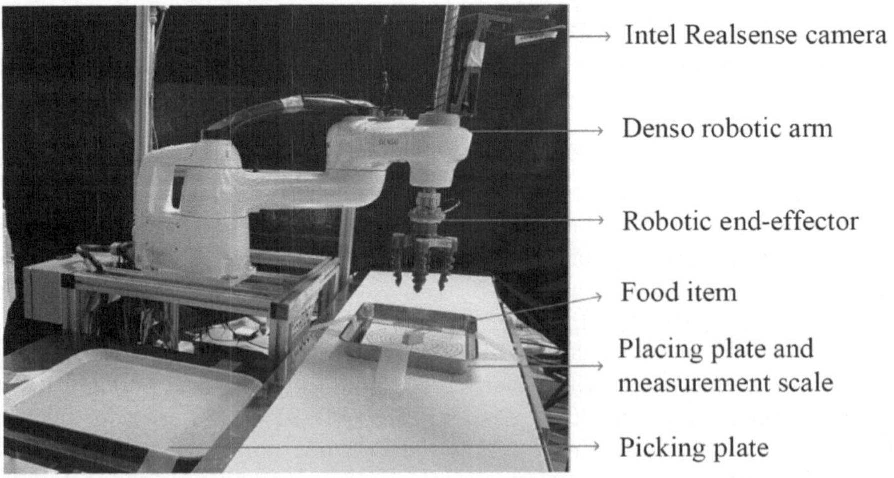

FIGURE 8.3
Experimental setup for the evaluation tests. Reproduced with permission under the CC-BY 4.0 license [58].

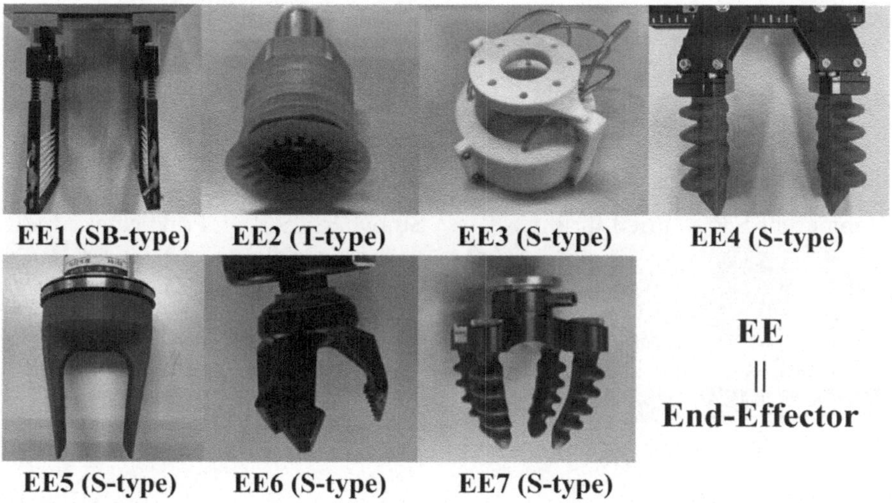

FIGURE 8.4
Selected robotic hands for the evaluation tests. Reproduced with permission under the CC-BY 4.0 license [58].

8.3.2 Tested robotic hands

Seven robotic hands were used for the handling experiments as shown in Figure 8.4. Five of them (EE2, EE4, EE5, EE6, EE7) are commercially available ones. EE1 and EE3 are under researched ones and were introduced previous in this book as the scooping-binding hand (EE1, **SB-type**, Section 2.3) and the circular membrane hand (EE3, **S-type**, Section 2.4). The commercialized robotic hands are a suction cup (EE2, **T-type**), two-fingered grippers (EE4 and EE5, **S-type**), a three-fingered hand (EE6, **S-type**), and a four-fingered gripper (EE7, **S-type**).

8.3.3 Tested food items

Based on the food categorization outlined in the previous chapter, 14 food items were selected for the handling experiments, as shown in Figure 8.5, to represent as many food categories as possible. To explore the feasibility of using food samples as substitutes for real foods in handling experiments, samples of the selected food items were also included. The handling-related physical properties of the real food items are presented in Table 8.1, with size referring to the dimension in the grasping direction. Weight and size measurements were taken from three real food items for each type, with pasta size determined by the plate dimension. Elasticity (Young's modulus) for all food items except pasta was measured using the method described in Section 6.2. Friction was measured as the maximum static friction coefficient using

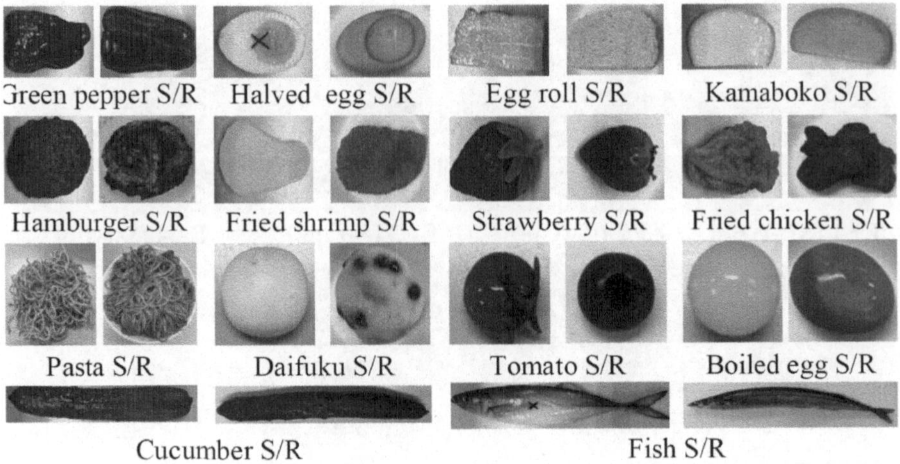

FIGURE 8.5
Selected food items for the evaluation tests. Reproduced with permission under the CC-BY 4.0 license [58].

TABLE 8.1
Handling-related physical properties of the 14 food items

Food items	Weight [g]	Size [mm]	Shape	Elasticity [kPa]	Friction
Green pepper	33∼38	46∼54	IR	166.00	1.35∼1.53
Halved egg	21∼23	37∼39	SR	43.47	0.32∼0.46
Egg roll	25∼26	28∼29	R	25.11	0.41∼0.59
Kamaboko	5∼7	21∼22	SR	52.27	1.42∼1.68
Hamburger	139∼141	60∼67	SR	15.00	0.34∼0.42
Fried shrimp	22∼24	40∼41	IR	384.40	0.50∼0.72
Strawberry	14∼18	25∼28	IR	29.55	0.57∼0.95
Fried chicken	25∼32	38∼45	IR	19.55	0.42∼0.64
Pasta	309∼310	200	IR	158.00	0.20∼0.30
Daifuku	55∼62	44∼46	SR	11.15	1.54∼1.69
Tomato	12∼13	24∼26	R	410.30	0.44∼0.65
Boiled egg	41∼42	38∼40	R	52.46	0.32∼0.46
Cucumber	89∼129	20∼28	SR	987.20	1.02∼1.23
Fish	103∼108	35∼37	SR	138.70	0.60∼0.71

Note: "R", "SR", and "IR" indicate regular, semi-regular, and irregular shapes, respectively. Reproduced with permission under the CC-BY 4.0 license [58].

a friction measuring machine (TL201 Tt, Trinity-Lab. Inc., Japan), with five measurements for each real food item type, excluding pasta. The elasticity and friction coefficient of pasta were obtained from [76] and [77], respectively.

8.3.4 Experimental protocols

The robotic manipulator and hand were controlled to grasp each food item and lift it to a height of approximately 200 mm. The item was then transported to a position about 150 mm above the designated placement point, followed by a downward movement to release the food item. The horizontal transport distance was approximately 540 mm, and the total duration of each pick-and-place cycle, including four pauses between each motion, was around 12 s. Air pressures for EE1, EE3, EE4, and EE7 were set at 60%, 80%, and 100% of their standard recommended pressures, respectively. For EE2, EE5, and EE6, which are actuated by vacuum pressure, experiments were conducted at three levels: approximately −40 kPa, −60 kPa, and −80 kPa. Five experimental trials were conducted for each food item, including both real foods and samples, resulting in a total of 7 (EEs) × 14 (foods) × 2 (real and sample) × 3 (pressures) × 5 (trials) = 2940 trials.

8.3.5 Evaluation results

Using robotic hand EE5 as an example, the evaluation results are displayed in Figure 8.6, where T1, T2, and T3 represent the three different air pressures used in the experiments. Successful trials (circles) indicate stable pickup and placement of the food item at the target location. Otherwise, trials were considered failures, categorized into three types: (1) grasping failure, where the item could not be picked up (triangles); (2) damage failure, where the item was damaged (squares); and (3) movement failure, where the item was dropped during transport (diamonds).

As shown in Figure 8.6, robotic hand EE5 succeeded in all trials for both the sample and real tomato but encountered 11 grasping failures with the real egg roll due to damage. Additional, EE5 failed in all trials involving the hamburger, as it was too large to fit within EE5's graspable area. Interestingly, Figure 8.6 also shows that EE5 succeeded in a few trials with real fish but failed in all trials with the fish sample due to the sample's heavier weight. This figure provides insights into how robotic hand EE5 performs with various food items.

Using the scoring method introduced in Section 8.2, scores were calculated for each robotic hand handling each food item. These results are visualized with a radar chart. Figure 8.7 displays the evaluation results of robotic hand EE5 handling 14 real food items. It is evident that EE5 was unable to handle kamaboko, hamburgers, daifuku, and fish due to various factors. For successfully handled items, EE5 demonstrated adaptability across a wide range of food properties, as shown in Figure 8.7o.

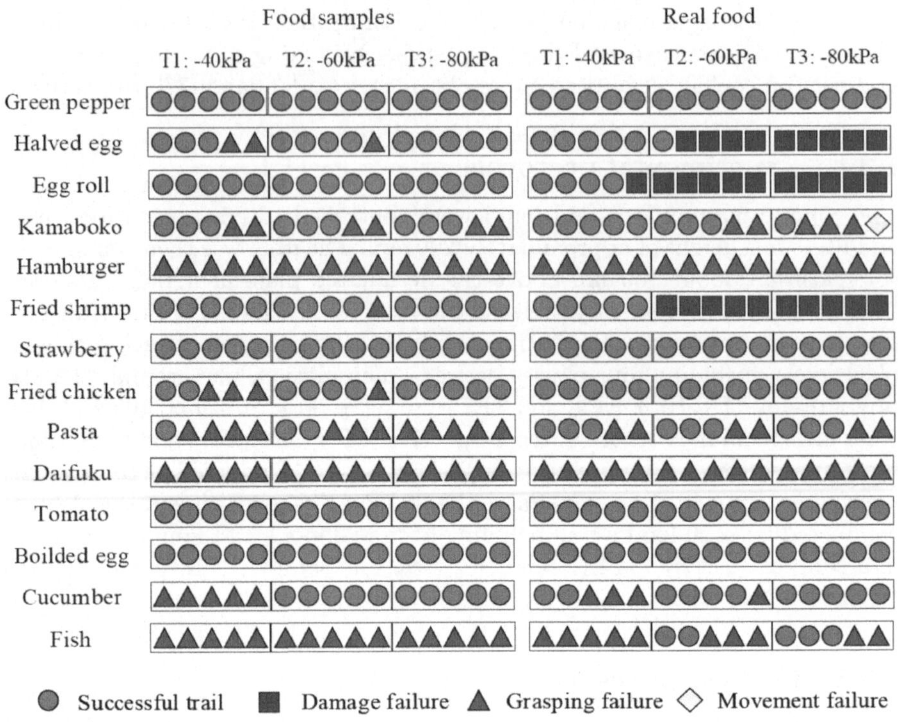

FIGURE 8.6
Evaluation results of robotic hand EE5. Reproduced with permission under the CC-BY 4.0 license [58].

Using radar charts, we can also compare the performance of different robotic hands for handling the same food item. Figure 8.8 presents the evaluation results of seven robotic hands handling fried shrimp. Robotic hands EE4, EE5, and EE6 performed well with this item. Additionally, EE1 and EE7 showed slightly better performance than EE4, EE5, and EE6 due to their greater adaptability to fragility. In contrast, EE2 and EE3 were unable to grasp the fried shrimp due to poor shape and size adaptability, respectively, as shown in Figures 8.8b and c, where the corresponding polygon vertices pass through the center of the radar charts.

An index was introduced to evaluate the overall performance of robotic hands, aiming to facilitate the selection of an appropriate robotic hand for a given food item [58]. This index was calculated as the product of all adaptability scores for a specific food item. Using the product rather than the sum highlights differences among robotic hands more effectively and clearly reflects a score of 0, indicating handling failure. Table 8.2 shows the calculated indices for seven robotic hands handling 14 food items. For a given food item, a higher index indicates better performance by the robotic hand, while identical indices

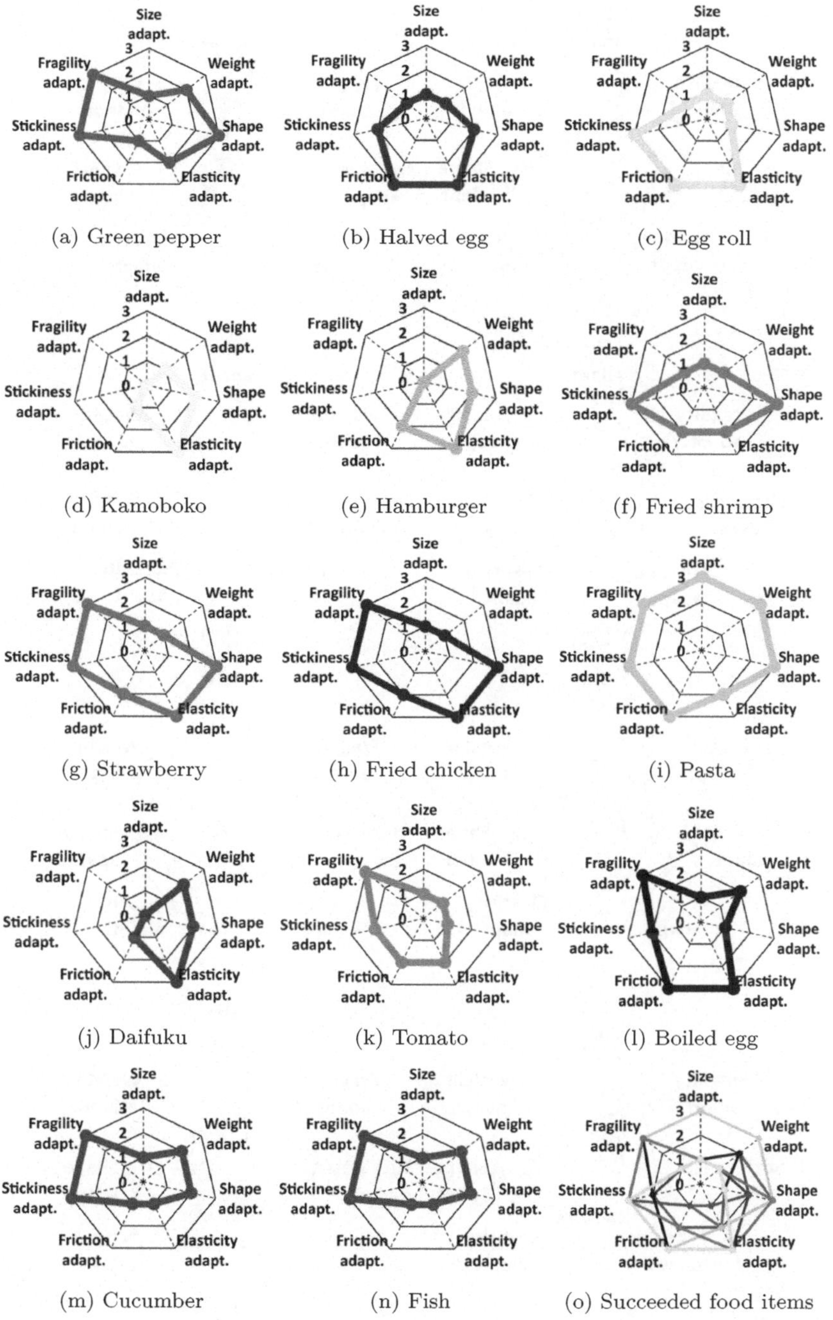

(a) Green pepper (b) Halved egg (c) Egg roll

(d) Kamoboko (e) Hamburger (f) Fried shrimp

(g) Strawberry (h) Fried chicken (i) Pasta

(j) Daifuku (k) Tomato (l) Boiled egg

(m) Cucumber (n) Fish (o) Succeeded food items

FIGURE 8.7

Evaluation results of EE5 visualizing in radar charts. The right bottom chart shows the overall results on handling all 14 food items. Reproduced with permission under the CC-BY 4.0 license [58].

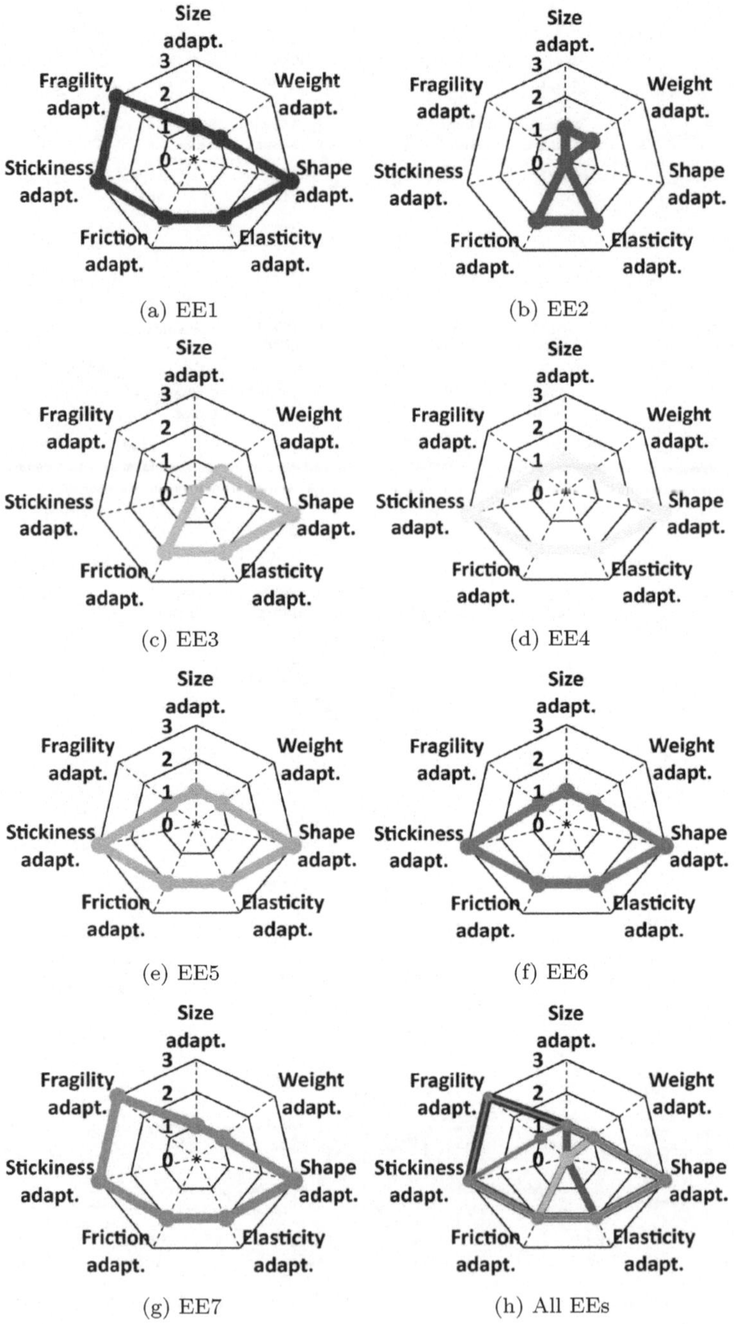

FIGURE 8.8
Evaluation results of seven robotic hands in handling the fried shrimp. The right bottom chart shows the comparison of all robotic hands. Reproduced with permission under the CC-BY 4.0 license [58].

TABLE 8.2
Evaluation indexes of the seven robotic hands for handling the 14 food items

Food item	EE1	EE2	EE3	EE4	EE5	EE6	EE7
Green pepper	108	0	0	0	108	72	108
Halved egg	108	0	0	108	36	108	108
Egg roll	0	0	0	27	27	54	0
Kamaboko	0	0	0	0	0	0	0
Hamburger	432	0	0	432	0	0	0
Fried shrimp	108	0	0	36	36	36	108
Strawberry	162	0	0	162	162	162	162
Fried chicken	162	0	0	0	162	162	162
Pasta	0	0	0	0	1458	1458	0
Daifuku	216	0	216	216	0	216	216
Tomato	24	0	0	0	24	24	24
Boiled egg	54	0	0	0	108	108	108
Cucumber	24	0	0	0	36	0	0
Fish	144	0	0	144	0	0	0

suggest comparable handling performance. An index of zero signifies that the robotic hand was unable to handle the food item. From Table 8.2, it is apparent that only two robotic hands were capable of handling pasta, hamburger, fish, and cucumber, while none succeeded with kamaboko. In contrast, multiple robotic hands successfully handled fried shrimp, halved egg, daifuku, and strawberry.

9

Case Study

Backcasting is a design methodology that begins with a clearly defined goal. In the context of food manipulation, this approach implies designing robotic hands based on the specific food items to be handled or the required manipulative operations. This chapter presents a case study on the backcasting development of a soft robotic hand optimized for high-speed handling of breadcrumb-coated oysters [78]. The robotic hand was successfully implemented in a production line for the automated packaging of breadcrumb-coated oysters into plastic trays.

9.1 Task Requirements

Deep-fried, breadcrumb-coated oysters are a popular dish in many countries, especially Japan. Commercially, raw breadcrumb-coated oysters are widely available, often sold frozen in supermarkets or online stores. In food processing facilities, these oysters are typically prepared by coating raw or defrosted oysters with breadcrumbs, placing them into plastic trays, and then transferring the packaged trays to large refrigerators. In most Japanese factories, the coating and transfer operations are automated using specialized conveyors; however, the packaging process remains largely manual.

In this case study, we collaborated with a major oyster manufacturer in Hiroshima Prefecture. According to the manufacturer, each production line requires four workers to manually place 16,000–18,000 breadcrumb-coated oysters into plastic trays per hour, translating to approximately 67–75 oysters/min per worker (illustrated in Figure 9.1). Automating this packaging step is in high demand not only to reduce labor costs but also to fully automate the production line. Complete automation provides significant advantages, including maintaining cold-chain integrity, reducing food contamination, and potentially achieving sterile production environments. For effective automation of the packaging operation, breadcrumb-coated oysters must be handled at high speed. The manufacturer requires an automation speed that matches or exceeds human capability, specifically 67–75 oysters/min.

DOI: 10.1201/9781003372431-9

FIGURE 9.1
Current process of manually packaging breadcrumb-coated oysters. Reproduced with permission under the CC BY license [78].

9.2 Tests Using Commercial Hand

To accomplish the task described above, the commercial soft gripper mGrip [66] was initially tested. The test setup included a parallel robot (FlexPicker, IRB 360, ABB), a camera system (scA1300-32gc, BSLER), and a belt conveyor, as shown in Figures 9.2a and b. The camera system detected oysters moving along the conveyor at a predetermined speed, enabling the parallel robot and attached gripper to grasp each oyster from the conveyor and transfer it to the designated tray area.

The gripper's trajectory is illustrated in Figure 9.2c. Upon approaching the pick position, set at 2 mm above the conveyor, a picking time of 0.1 s was configured to ensure proper gripper closure. After grasping the oyster, the gripper was directed to the designated placement position. The oyster's orientation was identified by the camera, and a rotation angle was calculated based on the difference between the oyster's detected orientation and the required orientation. This rotational adjustment occurred concurrently with the translational movement and was completed before placing the oyster in its designated tray slot. A placement time of 0.15 s was determined to be effective in ensuring precise positioning of the oyster within the tray slot.

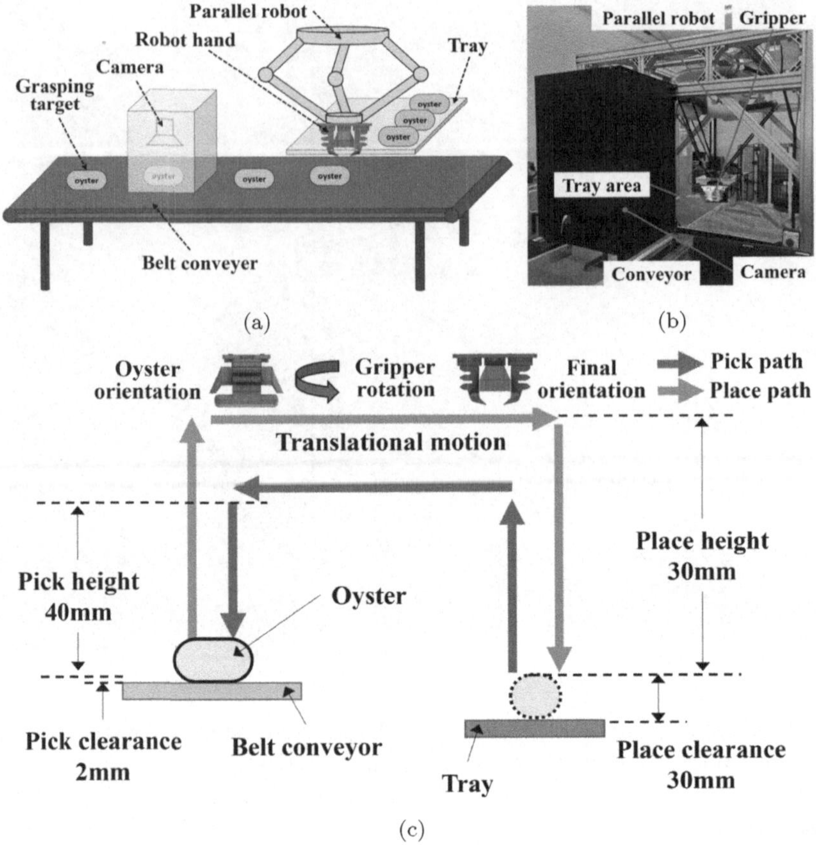

FIGURE 9.2
Experimental setup and hand motion for oyster handling: (a) system configuration, (b) actual robotic system, and (c) hand trajectory. Reproduced with permission under the CC BY license [78].

The tested items consisted of two types of oyster products, each weighing approximately 16 g and 40 g, as shown in Figure 9.3. A plastic tray commonly used in the production line was employed for the handling tests. The oysters were manually placed on the conveyor at three different orientations: approximately 0°, 45°, and 90°. An orientation of 0° indicates that the major axis of the oyster aligns with the conveyor's direction of movement and matches the required placement orientation within the tray. Consequently, oysters supplied at 45° and 90° required additional rotation to achieve proper alignment during placement. Air pressures of 21 kPa and 38 kPa were utilized in the experiments, as higher pressures caused excessive deformation of the breadcrumb coating on the oysters.

(a) (b)

FIGURE 9.3
Breadcrumb-coated oysters used in the experiments: (a) oyster weighted 16 g and (b) oyster weighted 40 g. Reproduced with permission under the CC BY license [78].

The experimental results are summarized in Table 9.1. The mGrip gripper was able to handle the 16 g oysters successfully at a rate of 60 oysters/min; however, at 70 oysters/min, it began to lose control, causing the oysters to be displaced. For the 40 g oysters, the mGrip gripper was unable to handle them even at a rate of 60 oysters/min. These findings suggest that the mGrip gripper cannot meet the required handling speed, indicating the need for a redesigned robotic hand.

TABLE 9.1
Experimental results of the mGrip gripper

Oyster	Speed	Pressure and oyster orientation					
		21 kPa			38 kPa		
	oysters/min	0°	45°	90°	0°	45°	90°
16 g	60	–	–	–	5/5	5/5	5/5
	70	–	–	–	5/5	4/5 ①	1/4 ①
40 g	60	0/5 ①②	–	–	0/1 ②	–	–

Note: The circled numbers indicate the reasons for the failed trials as ①: toss and ②: large deformation. "–" indicates the unconducted trials. Reproduced with permission under the CC BY license [78].

9.3 Challenge Analysis

To facilitate analysis, we examined two extreme scenarios, illustrated in Figure 9.4: (1) a gripper with very high stiffness, where the fingers remain rigid under acceleration, and (2) a gripper with very low stiffness, where the fingers exhibit significant bending under acceleration. In Scenario (1), as depicted in Figure 9.4a, a large impact occurs between the soft object and the rigid fingers, with the impact force increasing alongside acceleration. At a certain level of acceleration, this impact alone can damage the object. Conversely, in Scenario (2), shown in Figure 9.4b, the low stiffness of the gripper causes substantial swaying motion under acceleration, resulting in grasp failure and the object being dropped. These scenarios suggest that an optimal stiffness exists for the soft fingers, allowing a balance between minimizing impact forces and controlling sway under high acceleration.

To briefly estimate the damage caused by acceleration, the strain ε (%) under inertia force can be roughly estimated using the following equation:

$$\varepsilon = \frac{\sigma}{E} = \frac{F}{EA} = \frac{ma}{ELH} \times 100\%, \tag{9.1}$$

where E is Young's modulus of the breadcrumb-coated oyster, σ is stress, F is inertial force, a is acceleration, L and H are the length and height of the oyster, respectively. Here, we suppose that the contact area A is equal to the product of L and H which represents the best contact scenario. Table 9.2 listed the estimated strain of two kinds of oysters with weights of 16 g and 40 g under different accelerations, where "g" indicates the gravitation acceleration. The

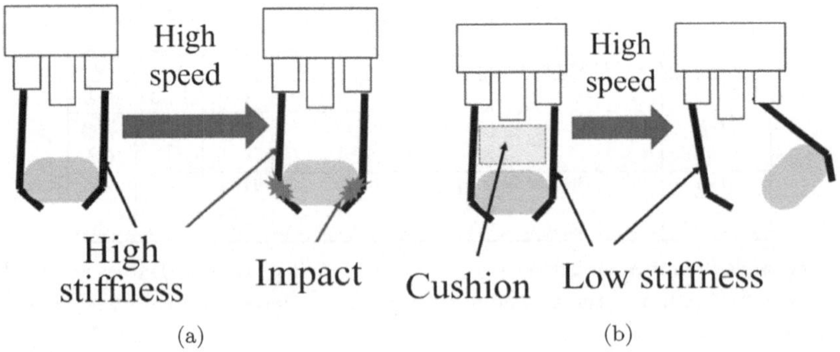

FIGURE 9.4
Two scenarios of soft gripper grasping soft object: (a) gripper with high stiffness and (b) gripper with low stiffness. Reproduced with permission under the CC BY license [78].

TABLE 9.2
Estimated oyster strain under different accelerations

Acceleration a	16 g oyster	40 g oyster
$1\,g$ ($9.8\,m/s^2$)	1.8%	2.9%
$5\,g$ ($49\,m/s^2$)	8.8%	14.4%
$10\,g$ ($98\,m/s^2$)	17.7%	28.8%
$20\,g$ ($196\,m/s^2$)	35.3%	57.6%

Note: Reproduced with permission under the CC BY license [78].

size of the oysters is averaged 63.6 mm×27.9 mm and 81.5 mm×33.4 mm for the 16 g and 40 g oysters, respectively. The Young's modulus was measured to be approximately 5 kPa. Table 9.2 suggests that an acceleration of $10\,g$ (the largest acceleration for the parallel robotic manipulator used in experiments) yielded approximately 20% strain under full contact conditions for the 40 g oyster. This strain level is unacceptable for maintaining product value. Therefore, the gripper must have appropriate stiffness to alleviate the impact of acceleration. In addition, equation (9.1) indicates that the contact area is important in reducing the strain. A larger contact area generates a smaller strain, which is usually desired to maintain product value. In addition, as illustrated in Figure 9.4b, a cushion between the two fingers is considered necessary to constrain the vertical upward motion of the target and provide soft contact to reduce damage.

9.4 Soft Containing Hand

Based on the above analysis, a soft robotic hand was proposed for handling breadcrumb-coated oysters under high acceleration. To achieve non-damaging and stable handling at high speeds, the hand design must offer optimal stiffness to mitigate inertial impact and provide a large contact area to minimize strain on the oyster. Additionally, the robotic hand must contain the oyster's motion in all directions without relying solely on friction-based grasping.

9.4.1 Hand design

As shown in Figure 9.5, the soft containing hand consists of a base, two soft fingers, and a soft center cushion. Figure 9.5b illustrates how the containing hand is designed to securely grasp and contain the oyster from all directions, with the containing space delineated by dotted lines in the cross-sectional view. The two soft fingers restrict lateral movement, while vertical motion is controlled by the soft cushion and the shovel-shaped ends of the fingers. Forward and backward movements are constrained by the "ears" of the soft

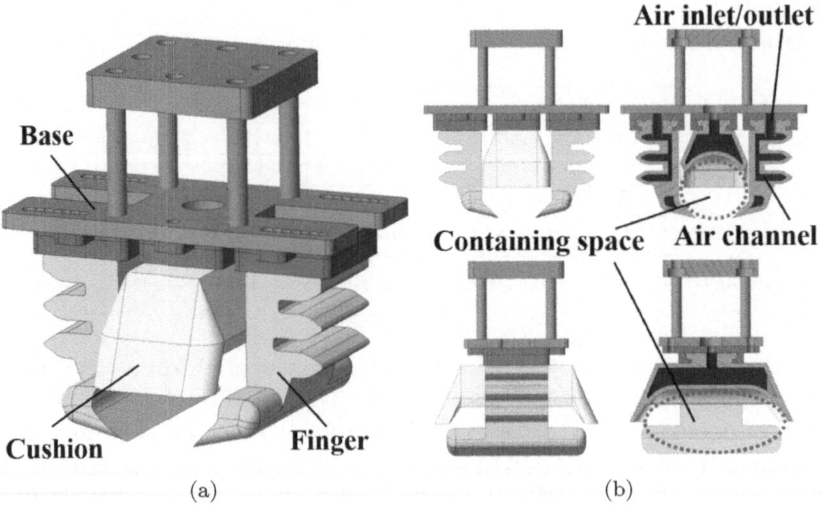

FIGURE 9.5
CAD design of the soft containing hand: (a) 3D isometric view and (b) front and side views along with cross-sectional views. The dotted circles indicate the containing space for holding the oyster. Reproduced with permission under the CC BY license [78].

cushion. This design maximizes the contact area between the hand and the oyster, effectively preventing the oyster from slipping once grasped.

9.4.2 Soft finger design

The CAD design of the soft finger is shown in Figure 9.6, with design parameters including b, c, d, e_1, e_2, f, and n, which will be used in subsequent analyses. Specifically, b denotes the neck wall thickness, c represents the neck height, d is the thickness of the bottom wall, e_1 and e_2 refer to the heights of the chamber troughs, f indicates the thickness of the shovel cavity wall (with $f = 0$ signifying the absence of a shovel cavity), and n represents the number of air chambers.

The finger features a shovel-shaped end that is significantly wider than the finger body, with a shovel angle of 135°. The total width of the shovel end was determined based on the size of the large oysters, which measures approximately 81.45 mm. This design choice maximizes the contact area during grasping. The shovel end incorporates a hollow structure, defined by the cavity wall thickness parameter f, to reduce inertial impact during high acceleration. Additionally, the chamber troughs are intentionally designed with different heights: the proximal chamber has a higher trough (e_1) than the distal chamber (e_2), providing the finger with the necessary stiffness to mitigate swaying motion during acceleration or deceleration.

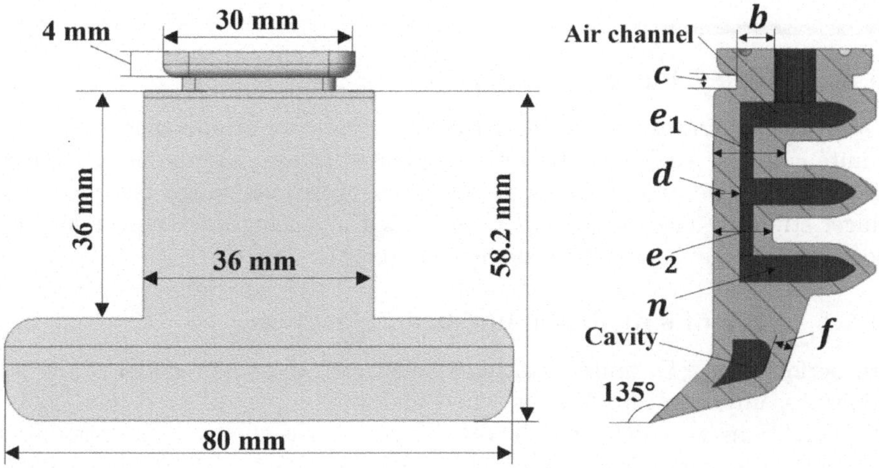

FIGURE 9.6
CAD design of the soft finger. Reproduced with permission under the CC BY license [78].

9.4.3 Soft cushion design

The soft cushion was designed to restrict the upward, forward, and backward movements of the oyster during high-speed motion. The CAD design of the soft cushion is shown in Figure 9.7. It features a hollow structure, ensuring both sufficient softness and low inertia. The overall size and circular concavity of the cushion were tailored to match the geometry of the large oysters, which have an average length of approximately 81.45 mm, width of 43.00 mm, and height of 33.35 mm.

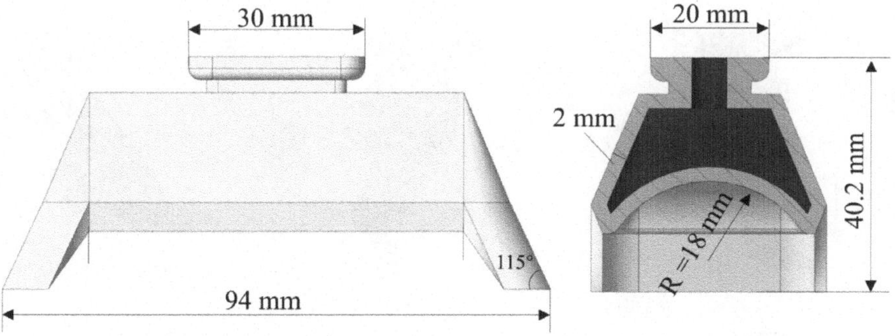

FIGURE 9.7
CAD design of the soft cushion. Reproduced with permission under the CC BY license [78].

9.5 Finite Element Analysis

Given the significant impact of actuator stiffness on gripper performance, a Finite Element Analysis (FEA) was conducted to examine the inertial effects on the soft hand and to assess the influence of various design parameters on finger stiffness. The simulations were carried out using the commercial FEA software Abaqus (Dassault Systèmes, France).

9.5.1 FEA of soft finger under acceleration

To perform the FE (finite element) simulations, the CAD model of the soft finger was imported into Abaqus, where material properties, mesh, and boundary conditions were defined. The soft finger was assigned a Young's modulus of 2.5 MPa, corresponding to a Shore hardness of A50. A 10-node quadratic tetrahedron hybrid element (C3D10H) was used to mesh the soft finger with a global mesh seed of 1 mm. For the interaction definition, the inner and outer surfaces of the finger chambers were treated as self-contact. The contact properties were modeled using "hard contact normal behavior" and "penalty tangential behavior", with a friction coefficient of 0.8. The top surface of the soft finger was fixed in space. The material density was set at 1080 kg/m^3 for silicone rubber [20]. A standard/implicit solver was employed, including geometrical nonlinearity to account for large deformation. The simulation was conducted in three steps, as shown in Figure 9.8: (1) applying a 1 g "gravity" load to the entire model along the vertical axis, (2) applying a vacuum pressure of −60 kPa to the inner surfaces of the air chambers, and (3) applying a 20 g acceleration as a "gravity" load along the horizontal axis.

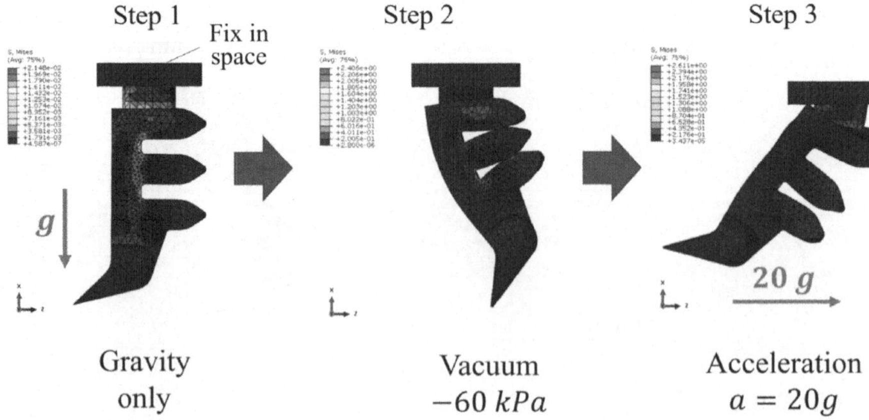

FIGURE 9.8
FE simulation of the soft finger. Reproduced with permission under the CC BY license [78].

The simulation results are shown in Figure 9.8, where the color represents the distribution of the Von Mises stress. The figure indicates that the soft finger does not experience significant deformation under gravity due to its low weight. Under a vacuum pressure of $-60\,\text{kPa}$, the soft finger bends toward the chamber side due to its hollow structure, representing the open state of the finger for grasping. When the vacuum pressure is maintained and the finger is subjected to a large acceleration of $20\,g$, it experiences substantial swaying from its vacuumed state. It is crucial to assess the swaying amplitude, as excessive motion could cause the gripper to collide with the object during grasping, potentially leading to grasp failure and object damage. As seen in Figure 9.8, the current finger stiffness is inadequate to maintain its shape during high-speed motion. These results highlight that inertial forces have a significant impact on the motion of the soft finger, suggesting that further investigation into the optimal stiffness of the finger is necessary.

9.5.2 FEA of soft fingers with different stiffnesses

FEA was conducted on different soft-finger designs to investigate the influence of finger stiffness. By modifying the initial design shown in Figure 9.8, nine new finger designs were developed to increase stiffness and reduce swaying motion under large acceleration. These nine designs are presented in Figure 9.9a. Modifications were made to the neck, bottom, shovel, and trough areas, with design parameters b, c, d, e_1, e_2, f, and n representing these changes. We hypothesize that increasing the neck thickness, shortening the neck, thickening the bottom, hollowing the shovel, thickening the trough, and reducing the number of chambers would result in higher finger stiffness. The same simulation steps shown in Figure 9.8 were performed on the nine new designs, and the final finger deformations are summarized in Figure 9.9b for comparison. The calculated swaying angles for each design are listed in Table 9.3.

The simulation results indicate that the stiffness of the soft finger can be adjusted by carefully designing the neck thickness and length, bottom thickness, shovel cavity, trough thickness, and the number of chambers, all without altering the overall size, material, or wall thickness of the air chambers. For example, introducing a cavity at the shovel end can reduce the mass of the soft finger by 14.1%, while simultaneously decreasing the swaying angle from 28.1° (design 2) to 15.6° (design 3). A thicker cavity wall (designs 6 and 8) results in a larger swaying angle compared to a thinner cavity wall (designs 5 and 7) due to the increased mass at the shovel end. Additionally, increasing the trough thickness by 2 mm reduces the swaying angle from 15.7° (design 6) to 7.6° (design 8). Designs 7 and 9 exhibited smaller swaying angles compared to other designs, with design 9 demonstrating higher stiffness due to its shorter length. To further assess the influence of gripper stiffness on object grasping, grasping simulations were conducted using finger designs 7 and 9.

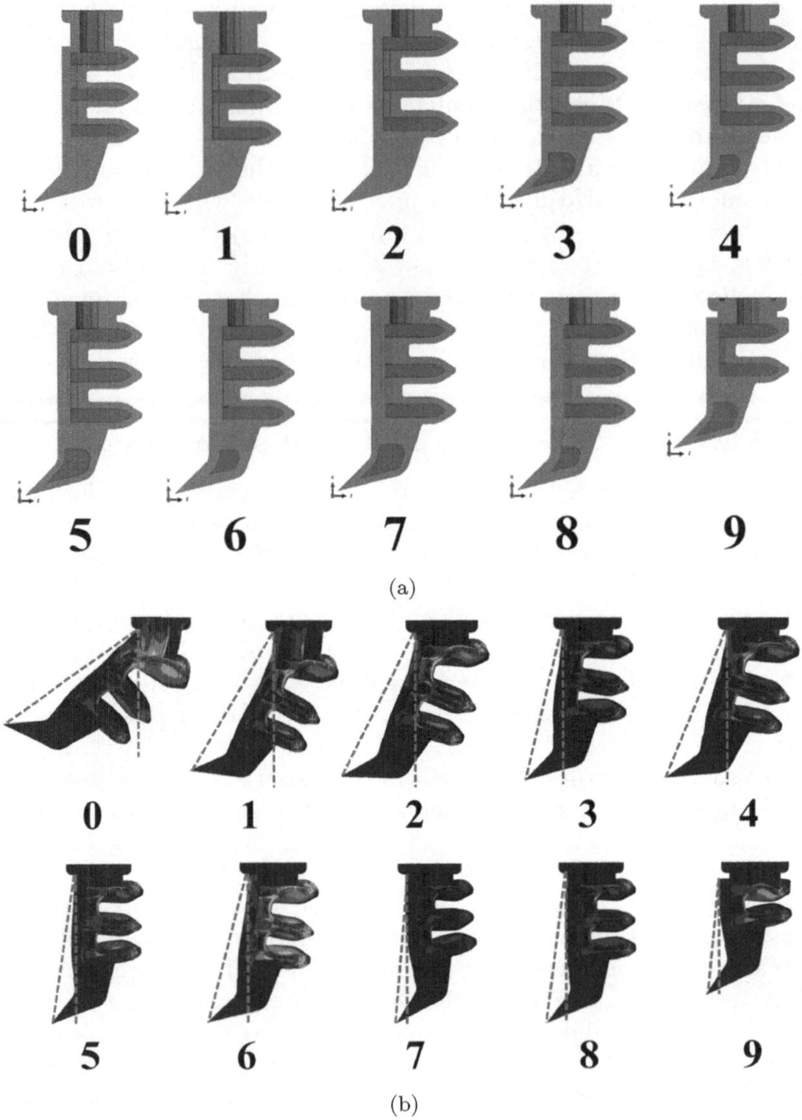

FIGURE 9.9
Different finger designs and their FEA results: (a) soft finger designs with different design parameters and (b) their FEA results under acceleration of 20 g where dotted lines indicate the swaying angles. Reproduced with permission under the CC BY license [78].

TABLE 9.3
Different soft finger designs and swaying angles under acceleration of $20\,g$ (design parameters of b, c, d, e_1, e_2, f, and n are described in Figure 9.6)

Design	Parameter	Angle
0	$n = 3$, $b = 3\,\text{mm}$, $c = 8\,\text{mm}$, $d = 3\,\text{mm}$ $e_1 = 8\,\text{mm}$, $e_2 = 8\,\text{mm}$, $f = 0$	54.5°
1	$n = 3$, $b = 5.5\,\text{mm}$, $c = 8\,\text{mm}$, $d = 3\,\text{mm}$ $e_1 = 8\,\text{mm}$, $e_2 = 8\,\text{mm}$, $f = 0$	31.6°
2	$n = 3$, $b = 5.5\,\text{mm}$, $c = 2.2\,\text{mm}$, $d = 3\,\text{mm}$ $e_1 = 8\,\text{mm}$, $e_2 = 8\,\text{mm}$, $f = 0$	28.1°
3	$n = 3$, $b = 5.5\,\text{mm}$, $c = 2.2\,\text{mm}$, $d = 3\,\text{mm}$ $e_1 = 8\,\text{mm}$, $e_2 = 8\,\text{mm}$, $f = 2\,\text{mm}$	15.6°
4	$n = 3$, $b = 5.5\,\text{mm}$, $c = 2.2\,\text{mm}$, $d = 3\,\text{mm}$ $e_1 = 8\,\text{mm}$, $e_2 = 8\,\text{mm}$, $f = 3\,\text{mm}$	23.2°
5	$n = 3$, $b = 5.5\,\text{mm}$, $c = 2.2\,\text{mm}$, $d = 4\,\text{mm}$ $e_1 = 9\,\text{mm}$, $e_2 = 9\,\text{mm}$, $f = 2\,\text{mm}$	11.6°
6	$n = 3$, $b = 5.5\,\text{mm}$, $c = 2.2\,\text{mm}$, $d = 4\,\text{mm}$ $e_1 = 9\,\text{mm}$, $e_2 = 9\,\text{mm}$, $f = 3\,\text{mm}$	15.7°
7	$n = 3$, $b = 5.5\,\text{mm}$, $c = 2.2\,\text{mm}$, $d = 4\,\text{mm}$ $e_1 = 11\,\text{mm}$, $e_2 = 9\,\text{mm}$, $f = 2\,\text{mm}$	6.2°
8	$n = 3$, $b = 5.5\,\text{mm}$, $c = 2.2\,\text{mm}$, $d = 4\,\text{mm}$ $e_1 = 11\,\text{mm}$, $e_2 = 9\,\text{mm}$, $f = 3\,\text{mm}$	7.6°
9	$n = 2$, $b = 5.5\,\text{mm}$, $c = 2.2\,\text{mm}$, $d = 4\,\text{mm}$ $e_1 = 9\,\text{mm}$, $f = 2\,\text{mm}$	5.6°

Note: Reproduced with permission under the CC BY license [78].

9.5.3 FEA of soft hand grasping soft object

Two hand models were constructed using finger designs 7 and 9, as shown in Figure 9.10a. To ensure a fair comparison, the containing spaces within each hand were kept identical, as indicated by the dotted lines, by adjusting the height of the cushion. To align with the conditions of the actual experiments using the parallel robot, the maximum acceleration was set to $10\,g$. The object was modeled as a capsule shape with a Young's modulus of 5 kPa, simulating the breadcrumb-coated oyster. Contact interactions between the soft fingers and the object, as well as between the center cushion and the object, were modeled as "surface-to-surface" contact. The contact properties were defined with 'hard contact normal behavior' and "penalty tangential behavior", with a friction coefficient of 0.8. The "ENCASTRE" boundary condition was applied to the top surfaces of the fingers and the cushion. All other settings were consistent with those used in the single-finger simulations. The simulations were conducted in three steps: (1) applying a pressure load of 60 kPa to the inner surfaces of the air chambers to grasp the object, (2) applying a $1\,g$ "gravity" load to the entire model, and (3) applying an acceleration of $10\,g$ as a "gravity" load to the entire model.

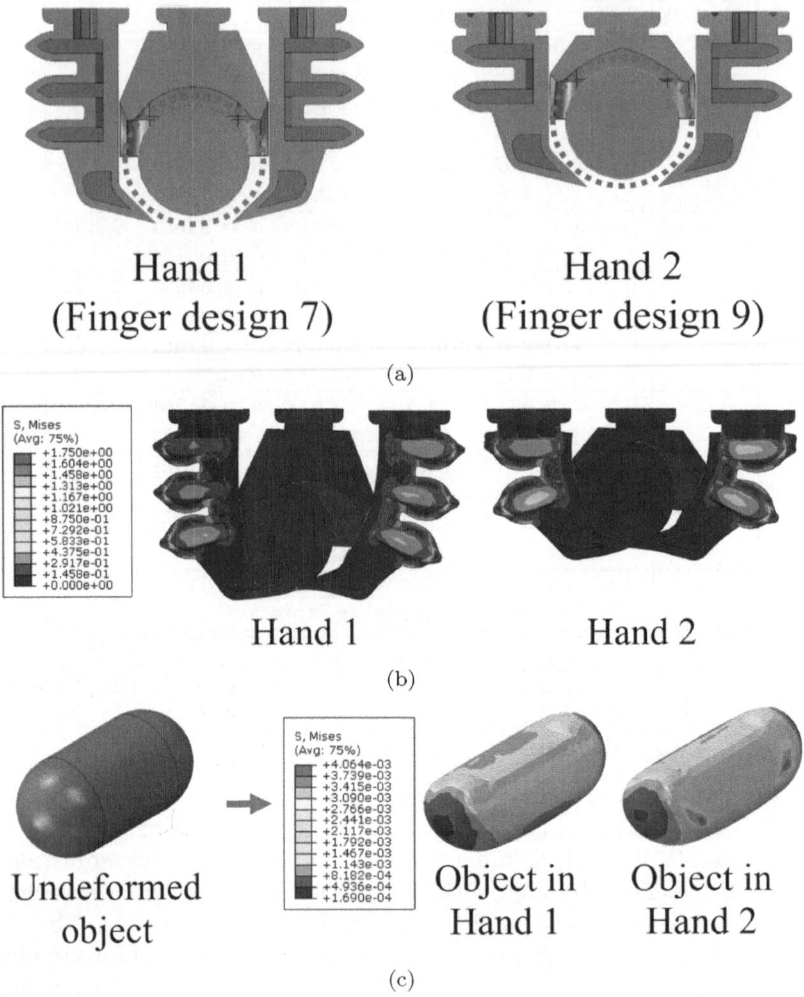

FIGURE 9.10
FEA of the soft hands grasping a soft object under acceleration: (a) section views of two hand models, ensuring same initial containing spaces; (b) simulation results of grasping a capsule-shaped soft object under an acceleration of $10\,g$; and (c) deformation and stress generated on the object. Reproduced with permission under the CC BY license [78].

Figure 9.10b presents the simulation results of the soft hands grasping the object under acceleration. The deformation and stress generated on the object during these simulations are shown in Figure 9.10c. The object grasped by Hand 1 experiences less stress and deformation compared to that grasped by Hand 2. Notably, Hand 2 has higher stiffness than Hand 1. These results suggest that an optimal stiffness for the soft hand exists when handling soft objects under high accelerations. To further validate this finding, both Hand 1 and Hand 2 were fabricated and subjected to experimental testing.

9.6 Hand Fabrication

To verify the hand design and guarantee its applicability in a food production line, we fabricated the soft hands using 3D printing technology and conducted durability and food compatibility tests.

9.6.1 Soft finger fabrication

The hand design incorporates hollow structures in the air chambers and a cavity in the shovel end. However, fabricating such structures using conventional casting methods is challenging. To address this, we employed 3D printing technology to fabricate the soft fingers. As shown in Figure 9.11a, we aimed to 3D print the soft finger in a single pass to ensure airtightness and durability. The 3D printer used for fabrication was the L320 (German RepRap), which is capable of 3D printing with liquid silicone material (SILASTIC 3D 3335) through liquid additive manufacturing (LAM) technology. The L320 deposits a layer of liquid silicone material through a mixer and nozzle, then uses a heat lamp to solidify each layer. Support materials are not compatible with the L320, and it can print inclined walls at angles of approximately 55° without the need for support structures. Therefore, the chamber design incorporated an inclination angle greater than 55° to ensure successful 3D printing.

To 3D print the shovel structure, a separate support part, fabricated in advance, was coated with detergent, as shown in Figure 9.11a (the thick line on the top surface of the support). The soft finger was then printed on top of the support component. After printing, the detergent facilitated easy separation of the soft finger from the support part. This approach utilizes a support component to enable the printing of a suspended structure. The detergent was chosen instead of a conventional silicone release agent to ensure the food compatibility of the fabricated soft finger, and it could be easily washed off with water.

The cavity inside the shovel structure presents another fabrication challenge, as it is a completely enclosed hollow region. To address this, we employed a low infill setting during 3D printing to create the cavity, as the

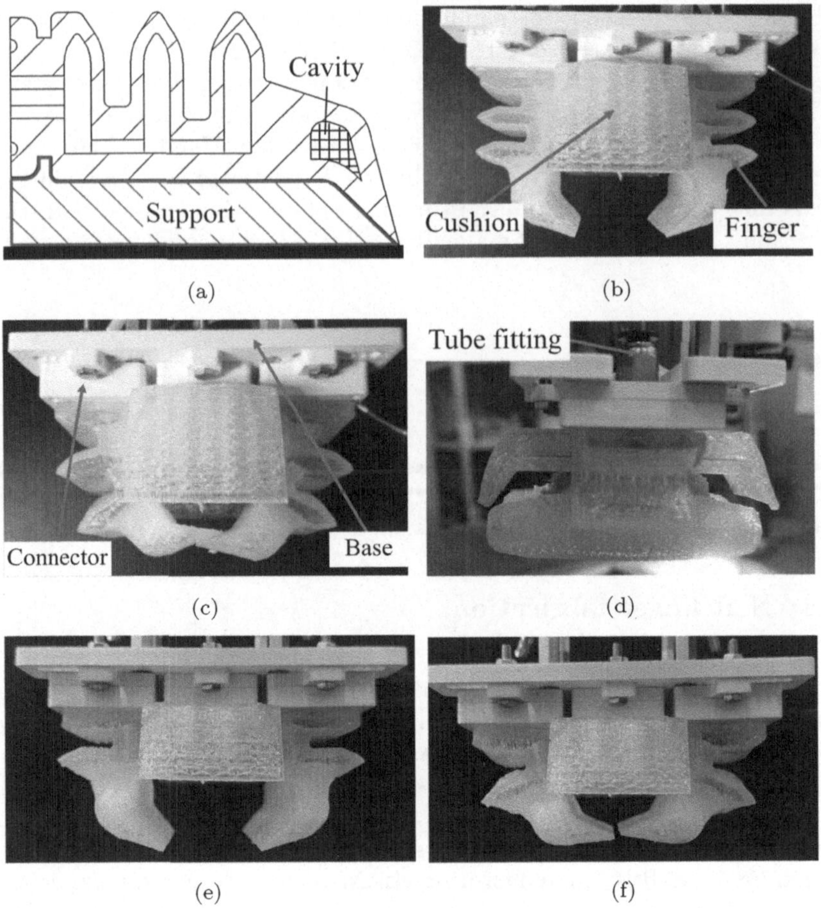

FIGURE 9.11
Soft hand fabrication: (a) cross-sectional drawing of the one-shot 3D printing method, (b) the assembled Hand 1 with three air chambers, (c) closed state of Hand 1, (d) soft hand containing an oyster, (e) the assembled Hand 2 with two air chambers, and (f) closed state of Hand 2. Reproduced with permission under the CC BY license [78].

primary purpose of the cavity is to reduce the weight of the shovel structure. An infill of 10% was used for fabricating the cavities. The fabrication of the center cushion followed a process similar to the one described above.

9.6.2 Hand assembly

The fabricated soft fingers and center cushions were assembled to construct two soft hands, as shown in Figures 9.11b and c. The assembly of the soft hands was achieved using connectors and a base (Figure 9.11c), which were 3D printed with an FDM 3D printer (Prusa i3 MK3S+, Prusa Research a.s.) using PLA material. Tube fittings (PC4-M5, NIHON PISCO Co., Ltd.) were employed to connect the soft hands to air tubes (Figure 9.11d). Upon pressurization, the soft hands close to form a space for grasping and containing, as shown in Figures 9.11c and d. Figure 9.11d demonstrates the soft hand containing a breadcrumb-coated oyster, confirming that the oyster is securely held within the grasping space and cannot escape from the hand.

For comparison, Hand 2, featuring shorter fingers (with two air chambers) and a shorter center cushion, was also fabricated. The initial and closed states of Hand 2 are shown in Figures 9.11e and f. Compared to Hand 1, which has three air chambers, Hand 2 exhibited greater bending stiffness due to its shorter length.

9.7 Durability Test

One concern with soft pneumatic grippers in practical applications is their durability, particularly regarding air leakage. This issue is especially pronounced when the gripper is fabricated using 3D printing, as the layered structures resulting from this technology often lead to weak bonding between layers, reducing durability. To address this concern, durability tests were conducted on the fabricated soft hands. Two types of durability tests were performed to cover various application scenarios: (1) a free bending test, where the gripper bends without contacting any object, and (2) a grasping test, where the gripper contacts a 3D-printed rigid sample of a breadcrumb-coated oyster. Experimental snapshots of the soft finger bending with and without contact with the sample are shown in Figures 9.12a and b. The 3D geometry of the sample was obtained using a 3D scanner (Artec Micro, Artec 3D), and the sample was 3D printed using PLA material.

The pneumatic system used for the durability tests is illustrated in Figure 9.12c. The air pressure was regulated through a setup consisting of an air compressor (JUN-AIR, USA), electro-pneumatic regulator (ITV2030, SMC), ejector (VUH07-66A, PISCO), microcomputer (Arduino Uno), MOSFET (SONONIA), and solenoid valve (VQ110-5M-M5, SMC). A pressure sensor (AP-C30(P), Keyence) was employed to monitor the air pressure applied to the soft fingers. The durability test was programmed to halt once the air pressure dropped from the initial 100 kPa to below 60 kPa. Black paper was placed beneath the soft fingers to collect any fragments that may have fallen off during the test. A web camera (Tinzzi x000wcl87p) was used to capture photographs every hour to document the test progress. The air pressures applied to the soft fingers were 100 kPa for inflation and −40 kPa for deflation.

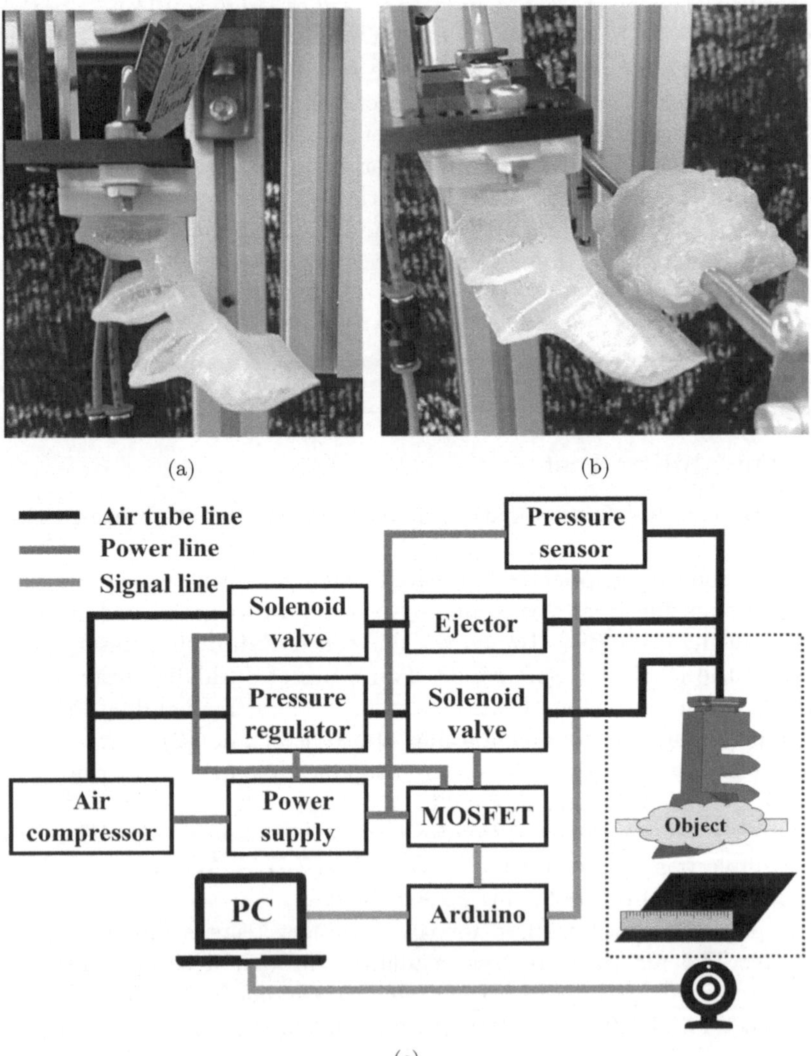

FIGURE 9.12
Durability test setup: different test scenarios of (a) without and (b) with contacting an object, and (c) the system diagram of the test setup. Reproduced with permission under the CC BY license [78].

The inflation and deflation durations were set at 0.3 s to simulate high-speed handling conditions.

The durability of the soft finger was tested under two conditions: with and without contact. In the no-contact scenario, the soft finger endured 2.15 million cycles, while in the contact scenario, it lasted 1.6 million cycles. In both cases, failure occurred at the chamber wall due to repeated inflation and stretching. The finger tested with contact showed better durability, as the presence of the target restrained the finger's flexure, reducing the stretch of the chamber wall. In contrast, without contact, the finger bent more, causing greater stretching of the chamber wall.

Notably, no broken pieces were observed in the test without contact, while 47 small broken fragments were observed during the contact test. The largest of these fragments measured approximately 0.5 mm. These small pieces were identified as burrs generated during the 3D printing process, highlighting the importance of post-processing before deployment. When considering the soft finger as a consumable component, its durability of over two million cycles would be sufficient, with a replacement interval of approximately two months—similar to the lifecycle of suction pads used in similar applications.

9.8 Food Compatibility Test

Given the critical importance of food safety in robotic food handling, ensuring the compatibility of the developed soft hands is essential. To assess this, food compatibility tests were conducted on the 3D-printed soft fingers through the hygiene test service provided by Japan Food Research Laboratories (JFRL [79]). The tests followed the 'Rubber Apparatus, Containers and Packaging Standard Testing based on Standards for Foods, Additives, etc.' The material tests focused on detecting the presence of cadmium and lead using the Alkali fusion method. Elution tests were conducted to measure phenol, formaldehyde, zinc, heavy metals, and evaporation residues (water).

To simulate both typical usage and steam cleaning scenarios, the tests were carried out at two temperature conditions: below and above 100°C. The results, summarized in Table 9.4, indicate that the gripper is safe for directly handling breadcrumb-coated oysters. To further ensure food safety, the soft fingers are regularly washed with clean water and sterilized at high temperatures to prevent the accumulation of stains.

TABLE 9.4
Results of food compatibility tests

Test item	<100°C	>100°C
Equipment and container packaging standards tests: rubber (excluding nursing equipment)	–	–
Material tests (alkali fusion method)	–	–
Cadmium and lead	–	–
Cadmium	Qualified	Qualified
Lead	Qualified	Qualified
Elution tests	–	–
Phenol	Qualified	Qualified
Formaldehyde	Qualified	Qualified
Zinc	Qualified	Qualified
Heavy metal	Qualified	Qualified
Evaporation residue (water)	Qualified	Qualified

Note: Reproduced with permission under the CC BY license [78].

9.9 Handling Tests for Implementation

Before implementing the robotic hand in the production line, handling tests were conducted using the experimental setup shown in Figure 9.2. The tests were performed at five different handling speeds, ranging from 60 to 100 oysters/min in 10 oysters/min increments, to evaluate the performance of the developed soft hands. Five consecutive pick-and-place trials were conducted for each handling speed, varying air pressures, oyster orientations, and both oyster products, across two soft hands.

9.9.1 Experimental results of Hand 1

Two air pressures, 69 kPa and 83 kPa, were used in the tests. These pressures were manually determined to generate a state of "boundary close" and "tight close" for Hand 1, respectively. The success rates for Hand 1 are summarized in Table 9.5. A success rate of 5/5 with a mark of "△" indicates that the handling was successful, but fragments of the breadcrumb coating were found. Trials marked with '–' were not conducted due to consecutive failures in the trials leading up to them, which only occurred at a handling speed of 100 oysters/min. Table 9.5 shows that Hand 1 successfully handled breadcrumb-coated oysters at speeds of 80 and 70 oysters/min for 16 g and 40 g oysters, respectively, at any oyster orientation. When oysters were supplied at an angle of 0°, the handling speed could be increased to 90 oysters/min for 16 g oysters

TABLE 9.5
Experimental results of Hand 1

Oyster	Speed [oysters/min]	Pressure and oyster orientation					
		69 kPa			83 kPa		
		0°	45°	90°	0°	45°	90°
16 g	60	5/5	5/5	5/5	5/5	5/5	5/5
	70	5/5	5/5	5/5	5/5	5/5	5/5
	80	5/5	5/5	5/5	5/5	5/5	5/5
	90	5/5	5/5 △	4/5 ②	5/5	5/5 △	5/5 △
	100	5/5 △	5/5 △	3/5 ②	–	–	4/5 ③
40 g	60	5/5	5/5	5/5	5/5	5/5	5/5
	70	5/5	5/5	5/5	5/5	5/5	5/5
	80	5/5	4/5 ②	2/5 ③	5/5	5/5 △	5/5 △
	90	5/5 △	5/5 △	4/4 △	5/5 △	5/5 △	4/5 ①
	100	5/5 △	4/5 △①	3/5 △①	2/5 △②	–	–

Note: The circled numbers indicate the reasons of failed trials as ①: misalignment in tray, ②: toss, and ③: large deformation. "△" and "−" indicate peelings of breadcrumb coating and unconducted trial, respectively.

and 80 oysters/min for 40 g oysters. If some peeling of the breadcrumbs was acceptable, the handling speed could reach 100 oysters/min for both oyster sizes. Handling failures due to increased speed and oyster angle included misalignment in the tray after placement, loss of grip resulting in the oyster being tossed, and unacceptably large deformation of the oyster, as indicated by the circled numbers.

9.9.2 Experimental results of Hand 2

Table 9.6 summarizes the experimental results for Hand 2. The applied air pressure was set to 76 kPa, which corresponds to the "boundary close" state. A higher pressure was required to reach the "boundary close" in Hand 2 due to its higher stiffness, which results from the shorter finger length. The results in Table 9.6 show that Hand 2 had a lower success rate compared to Hand 1. It was able to achieve a handling speed of 70 oysters/min for both oyster products at any oyster orientation. The speed could be extended to 80 oysters/min for the 16 g oysters if the oyster orientation was 0°. If some peeling of breadcrumbs was acceptable, the handling speed could reach 100 oysters/min for the 16 g oysters, but only 80 oysters/min for the 40 g oysters. When the oysters were

TABLE 9.6
Experimental results of Hand 2

Oyster	Speed [oyster/min]	Oyster orientation 0°	45°	90°
16 g	60	5/5	5/5	5/5
	70	5/5	5/5	5/5
	80	5/5	4/5 ①③	0/5 ②③
	90	5/5 △	2/5 ③	–
	100	5/5 △	1/5 ②③	–
40 g	60	5/5	5/5	5/5
	70	5/5	5/5	5/5
	80	5/5 △	5/5 △	2/5 ②③
	90	0/5 △③	–	–

Note: The circled numbers indicate the reasons of failed trials as ①: misalignment in tray, ②: toss, and ③: large deformation. "△" and "–" indicate peelings of breadcrumb coating and unconducted trial, respectively.

supplied at other orientations or at higher speeds, Hand 2 failed to handle the oysters, resulting in them being tossed away and damaged. This comparison indicates that Hand 1, with its lower stiffness, performs better than Hand 2, suggesting that an optimal finger stiffness exists for the task.

Additionally, by comparing the results in Tables 9.5 and 9.6 with those in Table 9.1, we found that the developed soft hands outperformed the mGrip gripper in handling breadcrumb-coated oysters.

9.9.3 Appearance of packaged oysters

Figure 9.13 presents snapshots of the packaged oysters handled by the three different soft hands. Handling failures are visible in the figure. For example, at a handling speed of 100 oysters/min, Hand 1 caused significant deformation in the 40 g oyster (first row, third column). Similar deformation failures are also observed with Hand 2 and the mGrip gripper at handling speeds of 90 oysters/min (second row, third column) and 60 oysters/min (third row, third column), respectively. Misalignment in the tray is evident with Hand 2 and the mGrip gripper when handling the 16 g oysters at handling speeds of 90 oysters/min (second row, second column) and 70 oysters/min (third row, second column), respectively.

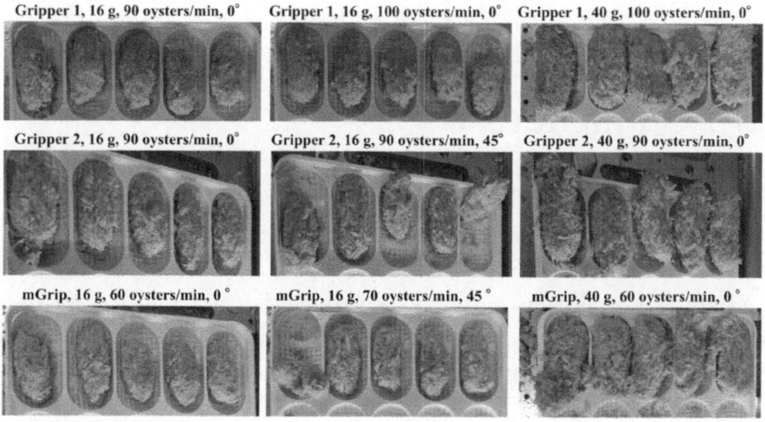

FIGURE 9.13
Snapshots of the packaged oysters using three different robotic hands. The first row: the Hand 1, the second row: Hand 2, and the third row: the mGrip gripper. Reproduced with permission under the CC BY license [78].

9.10 Discussions

Experimental results indicate that oyster orientation significantly affects the maximum handling speed, with proper orientation enhancing production efficiency. Poor orientation, on the other hand, can lead to misalignment during placement, tossing of the target during motion, and excessive deformation of the product. Table 9.5 suggests that the "boundary close" hand state, achieved with a pressure of 69 kPa, outperformed the 'tight close' state at 83 kPa. This supports the hypothesis that looser containment is more effective than tight grasping of soft objects, particularly in high-speed tasks. A hand state between "boundary close" and "tight close" may yield even better performance, warranting further investigation. Additionally, comparing the results in Tables 9.5 and 9.6, it was found that Hand 2, with shorter fingers and higher stiffness, performed worse than Hand 1, which had longer fingers and lower stiffness. The primary cause of failure was the excessive deformation of the oyster. This observation aligns with the FEA results in Figure 9.10 and the hypothesis in Figure 9.4, suggesting that higher stiffness of the soft finger does not necessarily improve performance when handling soft objects in high-speed tasks. In conclusion, there appears to be an optimal stiffness for soft fingers when handling soft objects in high-speed environments. As of the time of writing, two parallel robots equipped with the developed soft hands are in operation at a food factory in Hiroshima Prefecture, Japan. The performance of the robotic system in the production line can be seen in a YouTube video (https://www.youtube.com/watch?v=HTWyTYLS8rg).

10

Future Challenges

With the aging population and the resulting labor shortages, advancements in robotic technologies are expected to drive the increasing integration of robotic systems into the food industry, particularly for tasks traditionally performed by human laborers. This trend presents numerous opportunities for researchers and businesses in the fields of robotic manipulators, robotic end-effectors, computer science, artificial intelligence, and system integration.

10.1 Robotic Manipulator

In this book, we have not focused on robotic manipulators because industrial robotic manipulators are already widely used in the food industry [3], and there is limited research specifically targeting new robotic manipulators for food applications. However, this does not mean that there is no need to develop novel robotic manipulators for the food industry. In fact, most industrial robotic manipulators are not optimally suited for the specific needs of food-related tasks [80]. A key challenge is the high cost of current robotic systems, which poses a significant financial barrier for food manufacturers. As highlighted by Masey et al. [80], robotic manipulators designed for the food industry should meet specific requirements: easy-to-clean hygienic design, low cost, fast pick-and-place speed, safe operation alongside human workers, and ease of reprogramming. Recently, collaborative robots have become more prevalent in various applications due to their ability to operate alongside humans and ease of programming. However, challenges remain regarding their cost-effectiveness and operational speed. Additionally, soft robotic hands, which are increasingly used for handling delicate objects, often do not require high precision in positioning. Industrial robotic manipulators typically offer position accuracy ranging from 0.1 mm to 0.02 mm, which is crucial for automotive and electronics industries. However, for the food industry, an accuracy of several millimeters is usually sufficient. Therefore, the focus should be on improving operational speed and reducing costs in robotic manipulators for food applications.

DOI: 10.1201/9781003372431-10

10.2 Robotic End-Effector

Research on robotic end-effectors has been intensively conducted in recent years, including the studies introduced in this book. However, there are still relatively few commercialized end-effectors designed specifically for robotic food manipulation, and successful use cases in the food industry remain limited. To be successful, robotic end-effectors must not only meet hygienic design standards and operational speed requirements but also be low cost and easily integrable into existing robotic systems. Furthermore, a single end-effector should be versatile enough to handle a variety of food products, adapting to the rapid changes and updates commonly found in automation lines. Theoretical and experimental investigations are necessary to establish effective relationships between robotic end-effectors and food products. Additionally, as classified in Figure 8.1, end-effectors that grasp food from the top and side surfaces are well-studied and commercialized, but those that grasp from the bottom surface or from all surfaces have received little attention. However, such end-effectors are essential for handling a wide range of food products, particularly slippery, heavy, and low-profile items. It is expected that the demand for these types of end-effectors will increase as automation in the food industry advances.

10.3 Food Recognition

Food recognition in the RBP (robotic-based picking) scenario requires further investigation, particularly for the diverse range of food products and materials. Attention must be given not only to classification and recognition but also to accurately determining the grasping position and orientation. In the RBP context, these positions and orientations exist in 3D space, and any inaccuracies in recognition and determination could significantly impact grasping performance. To improve recognition accuracy, one promising approach is to leverage deep learning techniques. These methods, which learn representational features from large food datasets during the training process, have shown superior performance compared to traditional methods [81]. However, such datasets specifically tailored for RBP scenarios are not currently available and need to be developed.

10.4 Mapping between Foods and End-Effectors

To select appropriate robotic end-effectors for handling specific food products, it would be beneficial to have a mapping system that links food products to suitable robotic end-effectors. However, such a system is not yet available. In this book, we introduced classification methods for food products and robotic end-effectors based on handling-related food properties. However, quantitative data on some physical parameters, such as humidity, stickiness, and fragility, is still lacking. While evaluation experiments have been conducted on several robotic end-effectors, a comprehensive mapping between foods and end-effectors has not been addressed due to the limited number of tested food products and end-effectors. Additionally, the evaluation approach requires validation from peer researchers and professionals in the food industry. To enable the development of a mapping system, databases containing food properties and robotic end-effectors need to be created and made open source.

10.5 System Integration

Finally, from a systems perspective, a compact and portable design is ideal to accommodate the high-mix, low-volume production lines typical in the food industry. Moreover, food factories often lack dedicated system engineers, so robotic systems used in such environments must feature an intuitive interface, minimizing the need for complex maintenance, or providing automated maintenance solutions. With the advancement of the Fourth Industrial Revolution (4IR) and the widespread adoption of Internet of Things (IoT) and Cyber-Physical Systems (CPS) technologies [82], there is a growing opportunity to integrate these technologies into robotic systems in the food industry. This would enhance machine-to-machine communication, self-monitoring, and enable automatic system updates.

Bibliography

[1] W. Bruce Traill and Matthew Meulenberg. Innovation in the food industry. *Agribusiness*, 18(1):1–21, 2002.

[2] Robert Henry. Innovations in agriculture and food supply in response to the COVID-19 pandemic. *Molecular Plant*, 13(8):1095–1097, 2020.

[3] Farah Bader and Shahin Rahimifard. A methodology for the selection of industrial robots in food handling. *Innovative Food Science and Emerging Technologies*, 64, 2020.

[4] P.Y. Chua, T. Ilschner, and D.G. Caldwell. Robotic manipulation of food products – A review. *Industrial Robot: An International Journal*, 30(4): 345–354, 2003.

[5] T.K. Lien. 7 - Gripper technologies for food industry robots. In Darwin G. Caldwell, editor, *Robotics and Automation in the Food Industry*, Woodhead Publishing Series in Food Science, Technology and Nutrition, pages 143–170. Woodhead Publishing, 2013.

[6] Gualtiero Fantoni, Marco Santochi, Gino Dini, Kirsten Tracht, Bernd Scholz-Reiter, Juergen Fleischer, Terje Kristoffer Lien, Guenther Seliger, Gunther Reinhart, Joerg Franke, Hans Nørgaard Hansen, and Alexander Verl. Grasping devices and methods in automated production processes. *CIRP Annals*, 63(2):679–701, 2014.

[7] European Hygienic Engineering and Design Group (EHEDG), Doc. 8 Hygienic Design Principles, Third Edition, March 2018. URL: https://www.ehedg.org/guidelines-working-groups/guidelines/guidelines/detail/hygienic-design-principles.

[8] Christine Connolly. Abb high-speed picking robots establish themselves in food packaging. *Industrial Robot*, 34(4):281–284, 2007.

[9] Richard Vock, Alexander Dieckmann, Sebastian Ochmann, and Reinhard Klein. Fast template matching and pose estimation in 3D point clouds. *Computers & Graphics*, 79:36–45, 2019.

[10] Filip Ilievski, Aaron D. Mazzeo, Robert F. Shepherd, Xin Chen, and George M. Whitesides. Soft robotics for chemists. *Angewandte Chemie International Edition*, 50(8):1890–1895, 2011.

[11] Zhongkui Wang, Yuuki Torigoe, and Shinichi Hirai. A prestressed soft gripper: Design, modeling, fabrication, and tests for food handling. *IEEE Robotics and Automation Letters*, 2(4):1909–1916, 2017.

[12] Hisashi Iwamasa and Shinichi Hirai. Binding of food materials with a tension-sensitive elastic thread. In *2015 IEEE International Conference on Robotics and Automation (ICRA)*, pages 4298–4303, 2015.

[13] Zhongkui Wang, Damith Suresh Chathuranga, and Shinichi Hirai. 3D printed soft gripper for automatic lunch box packing. In *International Conference on Robotics and Biomimetics*, pages 503–508, 2016.

[14] Jun Shintake, Vito Cacucciolo, Dario Floreano, and Herbert Shea. Soft robotic grippers. *Advanced Materials*, 30(29):1707035, 2018.

[15] S. Davis, J.O. Gray, and G. Caldwell. An end effector based on the Bernoulli principle for handling sliced fruit and vegetables. *Robotics and Computer-Integrated Manufacturing*, 24:249–257, 2008.

[16] T.K. Lien and P.G.G. Davis. A novel gripper for limp materials based on lateral coanda ejectors. *CIRP Annals – Manufacturing Technology*, 57:33–36, 2008.

[17] Ethan W. Schaler, Donald Ruffatto, Paul Glick, Victor White, and Aaron Parness. An electrostatic gripper for flexible objects. In *2017 IEEE/RSJ International Conference on Intelligent Robots and Systems*, pages 1172–1179, 2017.

[18] Elliot W. Hawkes, David L. Christensen, Amy Kyungwon Han, Hao Jiang, and Mark R. Cutkosky. Grasping without squeezing: shear adhesion gripper with fibrillar thin film. In *2015 IEEE International Conference on Robotics and Automation*, pages 2305–2312, 2015.

[19] Elliot W. Hawkes, Hao Jiang, and Mark R. Cutkosky. Three-dimensional dynamic surface grasping with dry adhesion. *The International Journal of Robotics Research*, 35(8):943–958, 2015.

[20] Zhongkui Wang, Keung Or, and Shinichi Hirai. A dual-mode soft gripper for food packaging. *Robotics and Autonomous Systems*, 125, 2020.

[21] Takahiro Matsuno, Zhongkui Wang, and Shinichi Hirai. Gluing-free soft robotic hands for cake topping. In *IEEE/SICE International Symposium on System Integration*, 2024.

[22] Kevin C. Galloway, Kaitlyn P. Becker, Brennan Phillips, Jordan Kirby, Stephen Licht, Dan Tchernov, Robert J. Wood, and David F. Gruber. Soft robotic grippers for biological sampling on deep reefs. *Soft Robotics*, 3(1):23–33, 2016.

[23] Bianca S. Homberg, Robert K. Katzschmann, Mehmet R. Dogar, and Daniela Rus. Haptic identification of objects using a modular soft robotic gripper. In *2015 IEEE/RSJ International Conference on Intelligent Robots and Systems*, pages 1698–1705, 2015.

[24] Yufei Hao, Zheyuan Gong, Zhexin Xie, Shaoya Guan, Xingbang Yang, Ziyu Ren, Tianmiao Wang, and Li Wen. Universal soft pneumatic robotic gripper with variable effective length. In *2015 IEEE/RSJ International Conference on Intelligent Robots and Systems*, pages 6109–6114, 2016.

[25] Stefano Fusco, Mahmut Selman Sakar, Stephen Kennedy, Christian Peters, Salvador Pané, David Mooney, and Bradley J. Nelson. Self-folding mobile microrobots for biomedical applications. In *2014 IEEE International Conference on Robotics and Automation (ICRA)*, pages 3777–3782, 2014.

[26] Van Anh Ho, Hideyasu Yamashita, Zhongkui Wang, Shinichi Hirai, and Koji Shibuya. Wrin'tac: Tactile sensing system with wrinkle's morphological change. *IEEE Transactions on Industrial Informatics*, 13(5):2496–2506, 2017.

[27] Ali Hussein Qader, V.C. Agarwal, and Amer M. Ibrahim. Nonlinear behavior of continuous composite steel concrete beam with external prestress. *International Journal of Innovative Technology and Exploring Engineering*, 3(7):76–82, 2013.

[28] Yoshiyuki Kuriyama, Yuusuke Okino, Zhongkui Wang, and Shinichi Hirai. A wrapping gripper for packaging chopped and granular food materials. In *2019 2nd IEEE International Conference on Soft Robotics (RoboSoft)*, pages 114–119, 2019.

[29] B. Donald, L. Gariepy, and D. Rus. Distributed manipulation of multiple objects using ropes. In *IEEE International Conference on Robotics and Automation*, volume 1, pages 450–457, 2000.

[30] Zhongkui Wang, Haruki Furuta, Shinichi Hirai, and Sadao Kawamura. A scooping-binding robotic gripper for handling various food products. *Frontiers in Robotics and AI*, 8, 2021.

[31] Ryo Kanegae, Zhongkui Wang, and Shinichi Hirai. Easily fabricatable shell gripper for packaging multiple cucumbers simultaneously. In *2020 IEEE International Conference on Real-time Computing and Robotics (RCAR)*, pages 188–192, 2020.

[32] Zhongkui Wang, Ryo Kanegae, and Shinichi Hirai. Circular shell gripper for handling food products. *Soft Robotics*, 8(5):542–554, 2021.

[33] Yui Makiyama, Zhongkui Wang, and Shinichi Hirai. A pneumatic needle gripper for handling shredded food products. In *2020 IEEE International Conference on Real-time Computing and Robotics (RCAR)*, pages 183–187, 2020.

[34] Zhongkui Wang, Yui Makiyama, and Shinichi Hirai. A soft needle gripper capable of grasping and piercing for handling food materials. *Journal of Robotics and Mechatronics*, 33(4):935–943, 2021.

[35] Hiroki Aoyama, Zhongkui Wang, and Shinichi Hirai. Shell gripper inspired by human finger structure for automatically packaging agricultural product. In *2022 IEEE-RAS 21st International Conference on Humanoid Robots (Humanoids)*, pages 90–95, 2022.

[36] Kazuhiro Kosuge, Jina Lee, Junya Ichinose, and Yasuhisa Hirata. A novel grasping mechanism for flat-shaped objects inspired by lateral grasp. In *2008 2nd IEEE RAS & EMBS International Conference on Biomedical Robotics and Biomechatronics*, pages 282–288, 2008.

[37] Kazuhiro Kosuge, Yasuhisa Hirata, Jina Lee, Akira Kawamura, Koichi Hashimoto, Shingo Kagami, Yu Hayashi, Norio Yokoshima, Hiroshi Miyazawa, Ryosuke Teranaka, Yasutoshi Natsuizaka, Kazuhisa Sakai, Mizuo Koizumi, Junji Koyama, Naoki Kanayama, Shunichi Tezuka, and Torimitsu Hiroshi. Development of an automatic dishwashing robot system. In *2009 International Conference on Mechatronics and Automation*, pages 43–48, 2009.

[38] Yu Hayashi, Shingo Kagami, and Koichi Hashimoto. Visual measurement system for automated dishwashing. In *The proceedings of the Society of Instrument and Control Engineers Tohoku Chapter 249th Workshop*, 2009.

[39] Zhongkui Wang, Takao Hirata, Tomoharu Mori, Masaru Kawakami, Hidemitsu Furukawa, and Sadao Kawamura. A soft robotic hand based on bellows actuators for dishwashing automation. *IEEE Robotics and Automation Letters*, 6(2):2139–2146, 2021.

[40] Jyh-Hwa Tzou and Kuo L Su. The development of the restaurant service mobile robot with a laser positioning system. In *2008 27th Chinese Control Conference*, pages 662–666, 2008.

[41] Qingxiao Yu, Can Yuan, Zhuang Fu, and Zhao Yanzheng. Research of the localization of restaurant service robot. *International Journal of Advanced Robotic Systems*, 7(3):227–238, 2010.

[42] Qingxiao Yu, Can Yuan, Zhuang Fu, and Zhao Yanzheng. An autonomous restaurant service robot with high positioning accuracy. *Industrial Robot: An International Journal*, 39(3):271–281, 2012.

[43] Jia Guo, Xingyu Xiao, Panwang Pan, and Xiaotong Luo. A design of multi-vision localization and navigation service robot system. In *2017 12th International Conference on Computer Science and Education (ICCSE)*, pages 787–790, 2017.

[44] Takeshi Shimmura, Ryosuke Ichikara, Takashi Okuma, Hiroyuki Ito, Kei Okada, and Tomomi Nonaka. Service robot introduction to a restaurant enhances both labor productivity and service quality. In *13th CIRP Conference on Intelligent Computation in Manufacturing Engineering*, pages 589–594, 2020.

[45] Yudai Fukuzawa, Zhongkui Wang, Yoshiki Mori, and Sadao Kawamura. A robotic system capable of recognition, grasping, and suction for dishwashing automation. In *2021 27th International Conference on Mechatronics and Machine Vision in Practice (M2VIP)*, pages 369–374, 2021.

[46] J. Redmon, S. Divvala, R. Girshick, and A. Farhadi. You only look once: Unified, real-time object detection. In *2016 IEEE Conference on Computer Vision and Pattern Recognition*, pages 779–788, 2016.

[47] Jin Huat Low, Phone May Khin, Qian Qian Han, Haicheng Yao, Yee Seng Teoh, Yadan Zeng, Si Li, Jun Liu, Zhuangjian Liu, Pablo Valdiviay Alvarado, I-Ming Chen, Benjamin Chee Keong Tee, and Raye Chen Hua Yeow. Sensorized reconfigurable soft robotic gripper system for automated food handling. *IEEE/ASME Transactions on Mechatronics*, 27(5):3232–3243, 2022.

[48] Yitong Xue, Zhe Qiu, Huayan Zhang, Zhongkui Wang, and Shinichi Hirai. A synthetic dataset for robotic food handling system. In *2016 IEEE Conference on Computer Vision and Pattern Recognition*, pages 410–415, 2024.

[49] Z. Hertz and P. Lenard. On the contact of rigid elastic solids and on hardness. *Miscellaneous Papers*, Chapter VI:163–183, 1896.

[50] Zhongkui Wang, Kousuke Kadoma, and Shinichi Hirai. An indentation-type instrument for measuring soft tissue elasticity. In *Innovation in Medicine and Healthcare 2017*, pages 121–129, 2017.

[51] Zhongkui Wang and Shinichi Hirai. Modeling and parameter estimation of rheological objects for simultaneous reproduction of force and deformation. In *1st International Conference on Applied Bionics and Biomechanics*, 2010.

[52] Zhongkui Wang, Syogo Inoue, Yasutaka Hashimoto, and Sadao Kawamura. Measuring viscoelasticity and friction of tempuras for robotic handling. *Journal of Food Engineering*, 310:110707, 2021.

[53] Zhongkui Wang and Shinichi Hirai. Finite element modeling and physical property estimation of rheological food objects. *Journal of Food Research*, 1(1):48–67, 2012.

[54] Zhongkui Wang and Shinichi Hirai. Modeling and parameter identification of rheological object based on FE method and nonlinear optimization. In *The 2009 IEEE/RSJ International Conference on Intelligent Robots and Systems*, pages 1968–1973, 2009.

[55] Zhongkui Wang and Shinichi Hirai. Modeling and estimation of rheological properties of food products for manufacturing simulations. *Journal of Food Engineering*, 102(2):136–144, 2011.

[56] Helge A. Wurdemann, Vahid Aminzadeh, Jian S. Dai, John Reed, and Graham Purnell. Category-based food ordering processes. *Trends in Food Science & Technology*, 22(1):14–20, 2011.

[57] F. Erzincanli and J.M. Sharp. A classification system for robotic food handling. *Food Control*, 8(4):191–197, 1997.

[58] Zhe Qiu, Hannibal Paul, Zhongkui Wang, Shinichi Hirai, and Sadao Kawamura. An evaluation system of robotic end-effectors for food handling. *Foods*, 12(22):4062, 2023.

[59] Mitsui Chemicals, inc., Casting use: Polyurethane resins for castable materials, URL: https://jp.mitsuichemicals.com/en/service/product/casting/index.htm (Last access: March 16, 2025).

[60] SATO Sample, URL: https://310sample.cart.fc2.com/ (Last access: March 16, 2025).

[61] Wikipedia-croquette, URL: https://en.wikipedia.org/wiki/Croquette (Last access: March 16, 2025)

[62] Ebisu co., ltd., URL: https://ebisu.hp.gogo.jp/pc/ (last access: March 16, 2025)

[63] Wikipedia-hamburg steak, URL: https://en.wikipedia.org/wiki/Hamburg_steak (Last access: March 16, 2025)

[64] Hotty polymer: Super flexible type, URL: https://www.hotty.co.jp/product/hp-filament-superflexible/ (Last access: March 16, 2025)

[65] Foodly, the collaborative robot for food pick & place, URL: https://rt-net.jp/service/foodly-en/ (Last access: March 16, 2025)

[66] Soft Robotics, mGripAI: 3D Vision. Artificial Intelligence. Soft Grasping. URL: https://www.softroboticsinc.com/products/mgripai/ (Last access: March 16, 2025)

[67] Soft finger gripper – SFG, URL: https://www.softrobottech.com/web/en/category/13 (Last access: March 16, 2025)

[68] Softgripping: The modular design system for flexible gripping, URL: https://soft-gripping.com/ (Last access: March 16, 2025)

[69] Piab AB, piSOFTGRIP family, URL: https://www.piab.com/suction-cups-and-soft-grippers/soft-grippers/pisoftgrip-vacuum-driven-soft-gripper/pisoftgrip (Last access: March 16, 2025)

[70] Rochu soft robotic gripper group, Soft Robotics Grippers: Professional Solutions, URL: https://www.softroboticgripper.com/ (Last access: March 16, 2025)

[71] Furukawakikou Co., Ltd., SWITL, URL: https://www.furukawakikou.co.jp/switl/ (Last access: March 16, 2025)

[72] RightHand Robotics, Inc., RightPick Piece-Picking Solutions, URL: https://righthandrobotics.com/products (Last access: March 16, 2025)

[73] Shugen Ma, Liang Du, Erina Tsuchiya, and Mana Fuchimi. Paper-made grippers for soft food grasping. In *2020 17th International Conference on Ubiquitous Robots (UR)*, pages 362–367, 2020.

[74] A. Gafer, D. Heymans, D. Prattichizzo, and G. Salvietti. The quad-spatula gripper: A novel soft-rigid gripper for food handling. In *2020 3rd IEEE International Conference on Soft Robotics (RoboSoft)*, pages 39–45, 2020.

[75] Rosidah Sam and Samia Nefti. Design and feasibility tests of multi-functional gripper for handling variable shape of food products. In *2010 IEEE International Conference on Systems, Man and Cybernetics*, pages 1267–1272, 2010.

[76] Stefano Zardetto and Marco Dalla Rosa. Effect of extrusion process on properties of cooked, fresh egg pasta. *Journal of Food Engineering*, 92(1):70–77, 2009.

[77] Matthew Knuth and Chris Marone. Friction of sheared granular layers: Role of particle dimensionality, surface roughness, and material properties. *Geochemistry, Geophysics, Geosystems*, 8(3), 2007.

[78] Yang Zhang, Zhongkui Wang, Keigo Kutani, and Sadao Kawamura. A soft containing gripper for high-speed handling of breadcrumb-coated oysters. *Journal of Field Robotics*, 2024.

[79] JFRL (Japan Food Research Laboratories), URL: https://www.jfrl.or.jp/english (Last access: March 16, 2025)

[80] R.J. Moreno Masey, J.O. Gray, T.J. Dodd, and D.G. Caldwell. Guidelines for the design of low-cost robots for the food industry. *Industrial Robot*, 37(6):509–517, 2010.

[81] Lei Zhou, Chu Zhang, Fei Liu, Zhengjun Qiu, and Yong He. Application of deep learning in food: A review. *Comprehensive Reviews in Food Science and Food Safety*, 18(6):1793–1811, 2019.

[82] Zeashan Hameed Khan, Azfar Khalid, and Jamshed Iqbal. Towards realizing robotic potential in future intelligent food manufacturing systems. *Innovative Food Science & Emerging Technologies*, 48:11–24, 2018.

Index

G

H